Elementary Quantum Mechanics
in One Dimension

Elementary Quantum Mechanics in One Dimension

Robert Gilmore

The Johns Hopkins University Press

Baltimore & London

The Johns Hopkins University Press
2715 North Charles Street
Baltimore, Maryland 21218-4363
www.press.jhu.edu

ISBN 0-8018-8014-9
ISBN 0-8018-8015-7 (pbk.)

Library of Congress Control Number: 2004108544
A catalog record of this book is available from the British Library.

Contents

Part II Scattering

Part III Bound States

Part IV   Periodic Potentials

Preface

The small-scale physics of our modern world is governed by Quantum Mechanics. Many texts have been devoted to presenting this subject. Modulo style, they are all similar. Topics include Schrödinger's Equation, Laws of Quantum Mechanics, Operators and Commutation Relations, One-Dimensional Potentials, Separation of Variables, Theory of Angular Momentum, Hydrogen Atom, and so on.

Most texts do not treat solutions of one-dimensional problems in much detail. This is a pity for several reasons. First, most of the important physics (for spherically symmetric potentials, for example) comes from solving the one-dimensional radial equation. Second, elementary means exist to solve the one-dimensional equations. These involve multiplication of 2×2 matrices. Third, there is a rich interplay between the properties of similar binding potentials (eigenvalues/states), scattering potentials (resonances, probability densities), and periodic potentials (energy bands).

This book has been designed to serve as a complement to any of the standard texts commonly adopted for an undergraduate course in Quantum Mechanics. The standard material is not reviewed in the present work—this material is assumed from the outset. Instead, it focuses on the procedure required to "get results" and "develop understanding."

The method for constructing the 2×2 transfer matrix is presented in Part I. In this part we introduce the three different types of boundary conditions that are applied to the host of potentials that are discussed.

The three different types of boundary conditions are discussed extensively in the following three parts of this work: Part II—Scattering; Part III—Bound States; and Part IV—Periodic Potentials. The organization of topics in Parts II, III, and IV is very similar. Similar potentials are treated in each of these three parts. These treatments uncover a close relation between the properties of scattering states, bound states, and translationally invariant states in basically the same potential when subject to different boundary conditions. Such interrelations are either missing or opaque in the standard Quantum Mechanics texts. In each of these three cases, simple algorithms are presented to encourage readers to code and plot results themselves.

Material from this work has been used over the years in Quantum Mechanics courses taught several times at the Unviersity of South Florida and even more often at Drexel University.

This book is dedicated to my wife, Claire, and our children, Marc and Keith. To Claire, for taking me on a bicycle ride that changed my life. To Marc, for kicking me in the shins to affect my life. And to Keith, for appreciating the insights that have gone into it.

Part I

Foundations

1

Schrödinger's Equation

Schrödinger's equation is

$$\hat{H}\psi(\mathbf{x}, t) = i\hbar\frac{\partial\psi(\mathbf{x}, t)}{\partial t} \, . \tag{1.1}$$

Here $\psi(\mathbf{x}, t)$ is a wavefunction, and $\hat{H}$ is an operator obtained from the classical Hamiltonian describing a system. The classical Hamiltonian is a function of particle coordinates $\mathbf{x}_j, \mathbf{p}_j$, where $\mathbf{x}_j$ is the 3-vector describing the position of the jth particle and $\mathbf{p}_j$ is its momentum. The Hamiltonian operator $\hat{H}$ is obtained from the classical Hamiltonian H by making the substitutions

$$\mathbf{p}_j \to \frac{\hbar}{i}\, \nabla_j \, , \tag{1.2}$$

where ∇_j is the gradient operator acting on the coordinates of the jth particle, $h = 6.6255 \times 10^{-27}$ erg sec is Planck's constant, $\hbar = h/2\pi = 1.054 \times 10^{-27}$ erg sec, and $i = \sqrt{-1}$.

The Schrödinger equation (1.1) can be simplified by assuming a solution of the form $\psi(\mathbf{x}, t) = \Phi(\mathbf{x})e^{-iEt/\hbar}$. Then the explicit time dependence may be removed from (1.1) and the resulting time-independent equation is

$$\hat{H}\Phi(\mathbf{x}) = E\Phi(\mathbf{x}) \, . \tag{1.3}$$

The real constant E is interpreted as the energy of the system. The equations (1.1) and (1.3) are called the time-dependent and time-independent Schrödinger equations.

In this work, we will be particularly interested in the description of a single particle in one dimension. The classical Hamiltonian for a single particle in a potential $V(x)$ is

$$H = \frac{p^2}{2m} + V(x) \,. \tag{1.4}$$

Therefore, the Schrödinger equations we shall study are

$$\left[\frac{-\hbar^2}{2m} \frac{\partial^2}{\partial x^2} + V(x) \right] \psi(x,t) = i\hbar \frac{\partial \psi(x,t)}{\partial t} \,, \tag{1.5}$$

$$\left[\frac{-\hbar^2}{2m} \frac{d^2}{dx^2} + V(x) \right] \Phi(x) = E\Phi(x) \,. \tag{1.6}$$

2

Solutions in a Constant Potential

We will deal primarily with (1.6). To gain some familiarity with the time-independent Schrödinger equation (1.6), we shall solve it in a region of space in which the potential $V(x)$ is constant, $V(x) = V$. The equation may then be rewritten

$$\frac{d^2\Phi(x)}{dx^2} = \frac{-2m}{\hbar^2}(E - V)\Phi(x) \,. \tag{2.1}$$

Three types of solution may occur, depending on whether the term $-2m(E - V)/\hbar^2$ is negative, zero, or positive. Since under any condition we have a second-order differential equation to solve, there will always be two possible particular solutions.

Case A: $\frac{-2m}{\hbar^2}(E - V) = -k^2 < 0$,

$$\Phi_1(x) = e^{+ikx}, \qquad \Phi_2(x) = e^{-ikx} \,. \tag{2.2}$$

Case B: $\frac{-2m}{\hbar^2}(E - V) = 0$,

$$\Phi_1(x) = 1, \qquad \Phi_2(x) = x \,. \tag{2.3}$$

Case C: $\frac{-2m}{\hbar^2}(E - V) = +\kappa^2 > 0$,

$$\Phi_1(x) = e^{-\kappa x}, \qquad \Phi_2(x) = e^{+\kappa x} \,. \tag{2.4}$$

The most general solution of the Schrödinger equation in each of these three cases will be a complex linear superposition of the two particular solutions,

$$\Phi(x) = A\Phi_1(x) + B\Phi_2(x) \,. \tag{2.5}$$

5

The relationship beteween $E, V, k,$ and κ may be expressed in a transparent form as follows:

$$
\begin{array}{llll}
\text{Case A}: & \frac{\hbar^2 k^2}{2m} + V & = & E, \\
\text{Case B}: & V & = & E, \\
\text{Case C}: & \frac{-\hbar^2 \kappa^2}{2m} + V & = & E.
\end{array}
$$

In case A, the classical particle moves with a kinetic energy $KE = E - V = p^2/2m = (\hbar k)^2/2m$. Therefore, $\pm \hbar k$ may be interpreted as the momentum of a particle moving in a region of constant potential $V < E$.

We can make these considerations a little more precise by the following line of reasoning. When the Hamiltonian operator $\frac{-\hbar^2}{2m} \frac{d^2}{dx^2} + V(x)$ acts on the wavefunction $\Phi(x)$ it produces a multiple of the wavefunction (1.6). An equation of the form

$$
\text{(Operator)(Wavefunction)} = \text{(Number)} \times \text{(Wavefunction)} \qquad (2.6)
$$

is called an eigenvalue equation. (In (1.6), the number is the energy eigenvalue.) If we apply the momentum operator $\hat{p} = \frac{\hbar}{i} \frac{d}{dx}$ to the wavefunctions $\Phi_1(x) = e^{+ikx}, \Phi_2(x) = e^{-ikx}$, we should find the possible momentum states of the particle

$$
\begin{aligned}
\hat{p}\Phi_1(x) &= \frac{\hbar}{i}\frac{d}{dx}e^{+ikx} = (+\hbar k)e^{+ikx} = +\hbar k \Phi_1(x), \\
\hat{p}\Phi_2(x) &= \frac{\hbar}{i}\frac{d}{dx}e^{-ikx} = (-\hbar k)e^{-ikx} = -\hbar k \Phi_2(x).
\end{aligned}
\qquad (2.7)
$$

Therefore, $\Phi_1(x)$ represents a particle traveling in a region of constant potential $V < E$ with momentum $p = +\hbar k = +\sqrt{2m(E-V)}$ (i.e., to the right), while $\Phi_2(x)$ represents a particle traveling with momentum $p = -\hbar k = -\sqrt{2m(E-V)}$ (i.e., to the left).

A classical particle is forbidden to travel in a region in which $V > E$. This is reflected, in the quantum mechanical case, by the fact that the associated momenta are imaginary:

$$
\begin{aligned}
\hat{p}\Phi_1(x) &= \frac{\hbar}{i}\frac{d}{dx}e^{-\kappa x} = \frac{-\hbar\kappa}{i}e^{-\kappa x} = +i\sqrt{2m(V-E)}\Phi_1(x), \\
\hat{p}\Phi_2(x) &= \frac{\hbar}{i}\frac{d}{dx}e^{+\kappa x} = \frac{+\hbar\kappa}{i}e^{+\kappa x} = -i\sqrt{2m(V-E)}\Phi_2(x).
\end{aligned}
\qquad (2.8)
$$

The wavefunction $\Phi_1(x) = e^{-\kappa x}$ represents a solution of Schrödinger's equation that is exponentially decreasing toward the right, while $\Phi_2(x) = e^{+\kappa x}$ is exponentially increasing toward the right.

Case B, with $V = E$, is degenerate because the real momenta become zero in the limit $V \to E$ from below, or the imaginary momenta become zero in the limit $V \to E$ from above. Under this condition of degeneracy, mathematical theorems tell

Table 2.1 Solutions of Schrödinger's equation in a region with constant potential

Case	Two Independent Solutions	Eigenvalue of Momentum Operator	Definition of Parameters
A $V < E$	$\Phi_1 = e^{+ikx}$ $\Phi_2 = e^{-ikx}$	$+\hbar k$ $-\hbar k$	$+\frac{\hbar^2 k^2}{2m} + V = E$
B $V = E$	$\Phi_1 = 1$ $\Phi_2 = x$	0 Not an eigenfunction	$V = E$
C $V > E$	$\Phi_1 = e^{-\kappa x}$ $\Phi_2 = e^{+\kappa x}$	$+i\hbar\kappa$ $-i\hbar\kappa$	$-\frac{\hbar^2 k^2}{2m} + V = E$

us that at least one of the solutions must satisfy an eigenvalue equation, but the other solution need not:

$$\hat{p}\Phi_1(x) \;=\; \frac{\hbar}{i}\frac{d}{dx}1 \;=\; 0 \;=\; 0 \times \Phi_1(x)\,,$$

$$\hat{p}\Phi_2(x) \;=\; \frac{\hbar}{i}\frac{d}{dx}x \;=\; \frac{+\hbar}{i} \times 1 \;=\; \frac{\pm\hbar}{i} \times \Phi_1(x)\,.$$

The eigenvalue equation tells us that the corresponding momentum in case B is zero.
These results are summarized in Table 2.1.

3

Wavefunctions across a Boundary

In the previous chapter we have solved a very simple one-dimensional problem. In this chapter we shall solve a more complicated problem. We already know what a particle wavefunction looks like in a region in which the potential is constant. We now ask: What does a particle wavefunction look like if the potential has a constant value V_1 in one region of space (the line) and a different constant value V_2 in an adjacent region of space (Fig. 3.1)? We choose the break point between the two regions to be $x = a$. For the sake of concreteness, we temporarily assume the particle energy E is larger than either V_1 or V_2.

In region 1 the particle wavefunction $\Phi(x)$ is in general a linear superposition of the two specific solutions:

Region 1

$$x \leq a \qquad \Phi(x) = Ae^{ik_1x} + Be^{-ik_1x} ,$$

$$\frac{(\hbar k_1)^2}{2m} + V_1 = E . \tag{3.1}$$

Similarly, the wavefunction in region 2 is

Region 2

$$a \leq x \qquad \Phi(x) = Ce^{ik_2x} + De^{-ik_2x} ,$$

$$\frac{(\hbar k_2)^2}{2m} + V_2 = E . \tag{3.2}$$

In order to find a relationship between the wavefunctions for regions 1 and 2, we try to make the total wavefunction $\Phi(x)$ "as continuous as possible" across the boundary at $x = a$. Since Schrödinger's equation is a second-order differential equation, the solution in each region is characterized by two complex numbers [(A, B) in region 1; (C, D) in region 2]. Thus, we have two degrees of freedom to play with. That means we can choose the coefficients (A, B) and (C, D) so that the wavefunction and its first derivative are continuous at $x = a$:

Region 1 at $x = a$ Region 2 at $x = a$

$$
\begin{array}{rcl}
\Phi(a): & Ae^{ik_1a} + Be^{-ik_1a} & = & Ce^{ik_2a} + De^{-ik_2a} \\
\frac{d\Phi(a)}{dx}: & ik_1Ae^{ik_1a} - ik_1Be^{-ik_1a} & = & ik_2Ce^{ik_2a} - ik_2De^{-ik_2a}
\end{array}
\tag{3.3}
$$

This pair of simultaneous linear equations relating the coefficients (A, B) to the coefficients (C, D) can be handled in an elegant and simple way using matrix algebra:

$$
\begin{bmatrix} e^{ik_1a} & e^{-ik_1a} \\ ik_1e^{ik_1a} & -ik_1e^{-ik_1a} \end{bmatrix}
\begin{bmatrix} A \\ B \end{bmatrix}
=
\begin{bmatrix} e^{ik_2a} & e^{-ik_2a} \\ ik_2e^{ik_2a} & -ik_2e^{-ik_2a} \end{bmatrix}
\begin{bmatrix} C \\ D \end{bmatrix} .
\tag{3.4}
$$

In fact, the treatment becomes yet simpler if we first make it slightly more complicated by writing each 2×2 matrix in (3.4) as the product of two matrices, as follows:

$$
\begin{bmatrix} 1 & 1 \\ ik_1 & -ik_1 \end{bmatrix}
\begin{bmatrix} e^{ik_1a} & 0 \\ 0 & e^{-ik_1a} \end{bmatrix}
\begin{bmatrix} A \\ B \end{bmatrix}
=
$$
$$
\begin{bmatrix} 1 & 1 \\ ik_2 & -ik_2 \end{bmatrix}
\begin{bmatrix} e^{ik_2a} & 0 \\ 0 & e^{-ik_2a} \end{bmatrix}
\begin{bmatrix} C \\ D \end{bmatrix} .
\tag{3.5}
$$

In this form, the coefficients (C, D) for the wavefunction in region 2 may be related directly and simply to the coefficients (A, B) for the wavefunction in region 1 (or vice versa). The calculation is simple because it involves multiplication by matrix inverses.

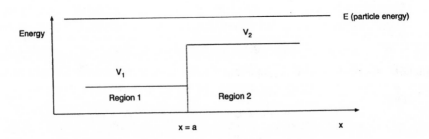

Fig. 3.1   The potential in the left-hand region is $V_1 < E$, and the potential in the right-hand region is $V_2 < E$. The wavefunction and its first derivative are matched at the boundary.

For example, to solve for (C, D) in terms of (A, B) we first multiply both sides of (3.5) by $\begin{bmatrix} 1 & 1 \\ ik_2 & -ik_2 \end{bmatrix}^{-1}$ on the left. Then we multiply by $\begin{bmatrix} e^{ik_2 a} & 0 \\ 0 & e^{-ik_2 a} \end{bmatrix}^{-1}$ on the left:

$$\begin{bmatrix} C \\ D \end{bmatrix} = \begin{bmatrix} e^{ik_2 a} & 0 \\ 0 & e^{-ik_2 a} \end{bmatrix}^{-1} \begin{bmatrix} 1 & 1 \\ ik_2 & -ik_2 \end{bmatrix}^{-1}$$

$$\times \begin{bmatrix} 1 & 1 \\ ik_1 & -ik_1 \end{bmatrix} \begin{bmatrix} e^{ik_1 a} & 0 \\ 0 & e^{-ik_1 a} \end{bmatrix} \begin{bmatrix} A \\ B \end{bmatrix}. \tag{3.6}$$

The coefficients (A, B) could just as easily have been solved for in terms of the coefficients (C, D) by a similar process

$$\begin{bmatrix} A \\ B \end{bmatrix} = \begin{bmatrix} e^{ik_1 a} & 0 \\ 0 & e^{-ik_1 a} \end{bmatrix}^{-1} \begin{bmatrix} 1 & 1 \\ ik_1 & -ik_1 \end{bmatrix}^{-1}$$

$$\times \begin{bmatrix} 1 & 1 \\ ik_2 & -ik_2 \end{bmatrix} \begin{bmatrix} e^{ik_2 a} & 0 \\ 0 & e^{-ik_2 a} \end{bmatrix} \begin{bmatrix} C \\ D \end{bmatrix}. \tag{3.7}$$

Equations (3.6) and (3.7) show that the coefficients (C, D) are related by a linear transformation to the coefficients (A, B).

For reasons that will become apparent at the end of chapter 4 (Fig. 4.2), the relation (3.7) is much preferable to the relation (3.6).

In deriving the relationship between the coefficients (A, B) in region 1 and the coefficients (C, D) in region 2, we have assumed $E > V_1, E > V_2$. We now relax this assumption. To discuss the general case it is only necessary to observe that in each of the three cases $V < E, V = E, V > E$, the wavefunction $\Phi(x)$ can be expressed as a linear superposition of the two particular solutions $\Phi_1(x), \Phi_2(x)$ given in (2.2)–(2.4)

$$\begin{aligned} \Phi(x) &= A\Phi_1(x) + B\Phi_2(x), \\ \Phi'(x) &= A\Phi_1'(x) + B\Phi_2'(x). \end{aligned} \tag{3.8}$$

The matrix relation between the wavefunction and its first derivative $\Phi(x), \Phi'(x)$ and the coefficients (A, B) is

$$\begin{bmatrix} \Phi(x) \\ \frac{d\Phi(x)}{dx} \end{bmatrix} = \begin{bmatrix} \Phi_1(x) & \Phi_2(x) \\ \Phi_1'(x) & \Phi_2'(x) \end{bmatrix} \begin{bmatrix} A \\ B \end{bmatrix}. \tag{3.9}$$

In detail, for the three possible cases we find:

Case A: $V < E$, $\qquad \frac{\hbar^2 k^2}{2m} + V = E$,

$$\begin{aligned} \Phi(x) &= Ae^{+ikx} + Be^{-ikx} \\ \frac{d\Phi(x)}{dx} &= ikAe^{+ikx} - ikBe^{-ikx} \end{aligned} \implies$$

$$\begin{bmatrix} \Phi(x) \\ \frac{d\Phi(x)}{dx} \end{bmatrix} = \begin{bmatrix} 1 & 1 \\ +ik & -ik \end{bmatrix} \begin{bmatrix} e^{+ikx} & 0 \\ 0 & e^{-ikx} \end{bmatrix} \begin{bmatrix} A \\ B \end{bmatrix}. \tag{3.10}$$

Case B: $V = E$,

$$
\begin{aligned}
\Phi(x) &= A \times 1 + B \times x \\
\frac{d\Phi(x)}{dx} &= 0 + B \times 1
\end{aligned}
\implies
$$

$$
\begin{bmatrix} \Phi(x) \\ \frac{d\Phi(x)}{dx} \end{bmatrix}
= \begin{bmatrix} 1 & 0 \\ 0 & 1 \end{bmatrix}
\begin{bmatrix} 1 & x \\ 0 & 1 \end{bmatrix}
\begin{bmatrix} A \\ B \end{bmatrix} .
\tag{3.11}
$$

Case C: $V > E$, $-\frac{\hbar^2 \kappa^2}{2m} + V = E$,

$$
\begin{aligned}
\Phi(x) &= A e^{-\kappa x} + B e^{+\kappa x} \\
\frac{d\Phi(x)}{dx} &= -\kappa A e^{-\kappa x} + \kappa B e^{+\kappa x}
\end{aligned}
\implies
$$

$$
\begin{bmatrix} \Phi(x) \\ \frac{d\Phi(x)}{dx} \end{bmatrix}
= \begin{bmatrix} 1 & 1 \\ -\kappa & +\kappa \end{bmatrix}
\begin{bmatrix} e^{-\kappa x} & 0 \\ 0 & e^{+\kappa x} \end{bmatrix}
\begin{bmatrix} A \\ B \end{bmatrix} .
\tag{3.12}
$$

Each of these equations can be written in the form

$$
\begin{bmatrix} \Phi(x) \\ \frac{d\Phi(x)}{dx} \end{bmatrix}
= K(V) E(V; x) \begin{bmatrix} A \\ B \end{bmatrix} .
\tag{3.13}
$$

The 2×2 matrices K, E as well as their inverses are collected in Table 3.1.

We return now to the problem of matching the wavefunction and its first derivative across a boundary, as illustrated in Fig. 3.1. Without making any assumptions about the relative values of E, V_1, V_2, the wavefunctions in regions 1 and 2 can be written as complex linear superpositions of the particular solutions $\Phi_1(x), \Phi_2(x)$ for the appropriate cases (Table 2.1). Matching the wavefunctions and their first derivatives at the boundary $x = a$ leads via (3.13) to the matrix equation

$$
K(V_1) E(V_1; a) \begin{bmatrix} A \\ B \end{bmatrix} = K(V_2) E(V_2; a) \begin{bmatrix} C \\ D \end{bmatrix} .
\tag{3.14}
$$

The expression for (A, B) in terms of (C, D) is then given very simply by

$$
\begin{bmatrix} A \\ B \end{bmatrix} = E^{-1}(V_1; a) K^{-1}(V_1) K(V_2) E(V_2; a) \begin{bmatrix} C \\ D \end{bmatrix} .
\tag{3.15}
$$

We have already encountered a special case of (3.15) in the case $E > V_1, E > V_2$ in (3.7).

Table 3.1   The 2×2 matrices $K(V)$ and $E(V; x)$, and their inverses, for the three cases: $E > V$, $E = V$, and $E < V$

	$E > V$	$E = V$	$E < V$
	$k = \sqrt{2m(E - V)/\hbar^2}$		$\kappa = \sqrt{2m(V - E)/\hbar^2}$
$K(V)$	$\begin{bmatrix} 1 & 1 \\ +ik & -ik \end{bmatrix}$	$\begin{bmatrix} 1 & 0 \\ 0 & 1 \end{bmatrix}$	$\begin{bmatrix} 1 & 1 \\ -\kappa & +\kappa \end{bmatrix}$
$K^{-1}(V)$	$\frac{1}{2}\begin{bmatrix} 1 & +\frac{1}{ik} \\ 1 & -\frac{1}{ik} \end{bmatrix}$	$\begin{bmatrix} 1 & 0 \\ 0 & 1 \end{bmatrix}$	$\frac{1}{2}\begin{bmatrix} 1 & -\frac{1}{\kappa} \\ 1 & +\frac{1}{\kappa} \end{bmatrix}$
$E(V; x)$	$\begin{bmatrix} e^{+ikx} & 0 \\ 0 & e^{-ikx} \end{bmatrix}$	$\begin{bmatrix} 1 & +x \\ 0 & 1 \end{bmatrix}$	$\begin{bmatrix} e^{-\kappa x} & 0 \\ 0 & e^{+\kappa x} \end{bmatrix}$
$E^{-1}(V; x)$	$\begin{bmatrix} e^{-ikx} & 0 \\ 0 & e^{+ikx} \end{bmatrix}$	$\begin{bmatrix} 1 & -x \\ 0 & 1 \end{bmatrix}$	$\begin{bmatrix} e^{+\kappa x} & 0 \\ 0 & e^{-\kappa x} \end{bmatrix}$

4

Piecewise Constant Potentials

In this chapter we solve Schrödinger's equation in one dimension with potentials more complicated than those used in the previous chapter. We consider here potentials that are constant in an interval

$$V(x) = V_j, \quad a_{j-1} < x < a_j . \tag{4.1}$$

Such a potential is illustrated in Fig. 4.1. The values of the potential at the breakpoints is unimportant as long as there are only a finite number of breakpoints.

4.1 TRANSFER MATRICES

Piecewise constant potentials can be treated by a simple extension of the methods developed in chapter 3. Instead of using different pairs of letters $(A, B), (C, D)$ for the particular solutions $\Phi_1(x), \Phi_2(x)$ in each region, we denote the general solution in region j by

$$\Phi(x) = A_j \Phi_1(x) + B_j \Phi_2(x), \quad a_{j-1} \leq x \leq a_j . \tag{4.2}$$

(Otherwise we might quickly run out of letters.) From (3.15) we know the matrix relation between the coefficients $(A, B) = (A_1, B_1)$ in region 1 and the coefficients $(C, D) = (A_2, B_2)$ in region 2 is

$$
\begin{aligned}
\begin{bmatrix} A_1 \\ B_1 \end{bmatrix} &= E^{-1}(V_1; a_1) K^{-1}(V_1) K(V_2) E(V_2; a_1) \begin{bmatrix} A_2 \\ B_2 \end{bmatrix} \\
&= T_{12} \begin{bmatrix} A_2 \\ B_2 \end{bmatrix} .
\end{aligned} \tag{4.3}
$$

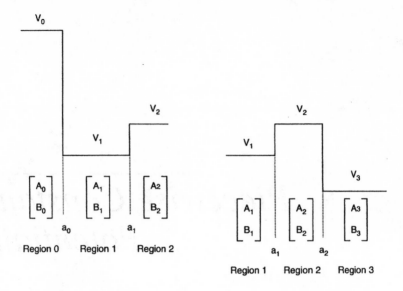

Fig. 4.1 The coefficients A_0, B_0 in the left-hand region 0 are related to the coefficients A_1, B_1 in Region 1 by a simple 2×2 transfer matrix. Similarly, the coefficients A_1, B_1 are related to A_2, B_2 by another simple transfer matrix. Thus, A_0, B_0 are related to A_2, B_2 by the product of simple 2×2 transfer matrices.

The break between regions 1 and 2 occurs at $x = a_1$. The 2×2 matrix T_{12} is called a transfer matrix because knowledge of the amplitudes A_2, B_2 can be transferred to knowledge of the amplitudes of A_1, B_1 with this matrix.

    Suppose now the asymptotic region on the left (region 0) and region 1 meet at breakpoint a_0 (Fig. 4.1). The coefficients (A_0, B_0) or (A_L, B_L) and (A_1, B_1) are related by an equation of the form (3.15):

$$
\begin{bmatrix} A_0 \\ B_0 \end{bmatrix} = E^{-1}(V_L; a_0) K^{-1}(V_L) K(V_1) E(V_1; a_0) \begin{bmatrix} A_1 \\ B_1 \end{bmatrix}
$$

$$
= T_{01} \begin{bmatrix} A_1 \\ B_1 \end{bmatrix}. \tag{4.4}
$$

Combining this with the matrix relation (4.3) yields an immediate linear relation between the coefficients (A_0, B_0) and (A_2, B_2)

$$
\begin{bmatrix} A_0 \\ B_0 \end{bmatrix} = T_{01} T_{12} \begin{bmatrix} A_2 \\ B_2 \end{bmatrix}. \tag{4.5}
$$

    Suppose now region 3 occurs to the right of region 2, the two regions meeting at breakpoint a_2 (Fig. 4.1). The coefficients (A_2, B_2) are related to (A_3, B_3) by an equation of the form (3.15):

$$\begin{bmatrix} A_2 \\ B_2 \end{bmatrix} = E^{-1}(V_2; a_2)K^{-1}(V_2)K(V_3)E(V_3; a_2) \begin{bmatrix} A_3 \\ B_3 \end{bmatrix}$$

$$= T_{23} \begin{bmatrix} A_3 \\ B_3 \end{bmatrix}. \tag{4.6}$$

Combining (4.6) with the linear relation between (A_1, B_1) and (A_2, B_2) given in (4.3) yields a linear relation between (A_1, B_1) and (A_3, B_3):

$$\begin{bmatrix} A_1 \\ B_1 \end{bmatrix} = T_{12}T_{23} \begin{bmatrix} A_3 \\ B_3 \end{bmatrix}. \tag{4.7}$$

For a piecewise constant potential with asymptotic constant value $V_0 = V_L$ on the left and $V_{N+1} = V_R$ on the right, there are $N + 1$ breakpoints $a_0, a_1, a_2, ..., a_N$, with a_j separating region j with constant potential V_j from region $(j + 1)$ with constant potential $V_{j+1}(j = 1, 2, ..., N)$. The linear relationship between (A_0, B_0) and (A_{N+1}, B_{N+1}), or (A_L, B_L) and (A_R, B_R), is easily seen to be

$$\begin{bmatrix} A_0 \\ B_0 \end{bmatrix} = T_{01}T_{12}T_{23}...T_{N-1,N}T_{N,N+1} \begin{bmatrix} A_{N+1} \\ B_{N+1} \end{bmatrix}$$

$$= T_{0,N+1} \begin{bmatrix} A_{N+1} \\ B_{N+1} \end{bmatrix}. \tag{4.8}$$

The individual matrices are

$$T_{j,j+1} = E^{-1}(V_j; a_j)K^{-1}(V_j)K(V_{j+1})E(V_{j+1}; a_j). \tag{4.9}$$

Notation: We will call $T_{0,N+1}$ (or simply T) the transfer matrix for the problem of a piecewise constant potential with asymptotic constant values V_0 (or V_L) on the left and V_{N+1} (or V_R) on the right.

The transfer matrix $T_{0,N+1}$ is obtained as a product of 2×2 matrices. There are four 2×2 matrices at each breakpoint (the product at a_j is $T_{j,j+1}$, given in (4.9)), so that a problem involving N piecewise constant potentials between the asymptotic potentials V_L on the left and V_R on the right, defined by $N + 1$ breakpoints, involves multiplying together $4(N + 1)$ 2×2 matrices.

It is useful to carry out whatever simplifications are possible before the actual computations are performed. In the present case a great simplification is possible. Consider the product of two successive transfer matrices

$$T_{j-1,j}T_{j,j+1} = E^{-1}(V_{j-1}; a_{j-1})K^{-1}(V_{j-1})$$

$$\tag{4.10}$$

$$\times \underbrace{K(V_j)E(V_j; a_{j-1})E^{-1}(V_j; a_j)K^{-1}(V_j)} \times K(V_{j+1})E(V_{j+1}; a_j).$$

The four interior matrices, which are underlined, refer only to region j, where the potential has constant value V_j. The product of these four matrices can easily be

Table 4.1   Real 2×2 matrices $M(V; \delta)$ for the three cases $E > V$, $E = V$, $E < V$

$$M(V_j, \delta_j) = K(V_j)E(V_j; a_j)E^{-1}(V_j; a_{j-1})K^{-1}(V_j)$$

Case A	Case B	Case C
$E > V$ $k = \sqrt{2m(E-V)/\hbar^2}$	$E = V$	$E < V$ $\kappa = \sqrt{2m(V-E)/\hbar^2}$
$\begin{bmatrix} \cos k\delta & -k^{-1}\sin k\delta \\ +k\sin k\delta & \cos k\delta \end{bmatrix}$	$\begin{bmatrix} 1 & -\delta \\ 0 & 1 \end{bmatrix}$	$\begin{bmatrix} \cosh\kappa\delta & -\kappa^{-1}\sinh\kappa\delta \\ -\kappa\sinh\kappa\delta & \cosh\kappa\delta \end{bmatrix}$
	$\delta = a_{j+1} - a_j$	$\cosh x = \frac{1}{2}(e^{+x} + e^{-x})$ $\sinh x = \frac{1}{2}(e^{+x} - e^{-x})$

computed. The matrix product

$$M(V_j, \delta_j) = K(V_j)E(V_j; a_{j-1})E^{-1}(V_j; a_j)K^{-1}(V_j) , \qquad (4.11)$$

($\delta_j = a_j - a_{j-1}$), is given in Table 4.1 for the three cases $E > V$, $E = V$, $E < V$. The matrix $M_j = M(V_j, \delta_j)$ depends only on the potential V_j in region j and the width δ_j of region j, as well as the particle energy E. Further, this matrix is always real and has determinant $+1$.

The computation of the transfer matrix simplifies to

$$T_{0,N+1} = E^{-1}(V_0; a_0)K^{-1}(V_0) \times M_1 M_2 \ldots M_N \times K(V_{N+1})E(V_{N+1}; a_N) . \qquad (4.12)$$

As a result, the computation of the transfer matrix involves the product of $N + 2 \times 2 = N + 4$ instead of $4(N+1) = 4N + 4$ matrices: two matrices each for the left and right asymptotic regions, and the N real 2×2 matrices M_j for the N interior regions.

4.2   COMPUTATIONAL ALGORITHM

We show in Fig. 4.2 a very simple algorithm for constructing the transfer matrix merely by inspecting a piecewise constant potential. This algorithm would not have been as direct had we adopted the solution (3.6) instead of (3.7).

In summary, the algorithm for computing the transfer matrices that we will use is as follows. Equation (4.12) relates the amplitudes $\begin{bmatrix} A_L \\ B_L \end{bmatrix}$ in the asymptotic left-hand region with the amplitudes $\begin{bmatrix} A_R \\ B_R \end{bmatrix}$ in the asymptotic right-hand region. It is

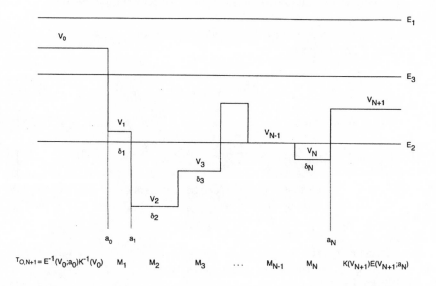

Fig. 4.2 The transfer matrix for the interior pieces of a piecewise constant potential is very simply constructed by inspection. We associate a simple 2×2 real transfer matrix with each piece of the potential and then simply multiply them in the order in which they occur.

often useful to absorb the diagonal matrix elements of $E(V_L; a_0)$ into the amplitudes $\begin{bmatrix} A_L \\ B_L \end{bmatrix}$ and the diagonal matrix elements of $E(V_R; a_N)$ into the amplitudes $\begin{bmatrix} A_R \\ B_R \end{bmatrix}$. Then

$$E(V_L; a_0)\begin{bmatrix} A_L \\ B_L \end{bmatrix} = K^{-1}(V_L)M_1 \ldots M_N K(V_R)E(V_R; a_N)\begin{bmatrix} A_R \\ B_R \end{bmatrix},$$

$$\begin{bmatrix} A_L' \\ B_L' \end{bmatrix} = K^{-1}(V_L)\left\{ \prod_{j=1}^{N} M(V_j; \delta_j) \right\} K(V_R)\begin{bmatrix} A_R' \\ B_R' \end{bmatrix}. \quad (4.13)$$

The first step is the computation of the 2×2 real unimodular matrices $M(V_j; \delta_j)$ for each of the N intermediate piecewise constant potentials of energy V_j and width δ_j. These matrices can be written down by inspection for any energy E. The product of these N matrices is then computed in the order in which the potentials appear. We call the product M:

$$M = M(V_1; \delta_1) \ldots M(V_N; \delta_N) = \prod_{j=1}^{N} M(V_j; \delta_j) = \begin{bmatrix} m_{11}(E) & m_{12}(E) \\ m_{21}(E) & m_{22}(E) \end{bmatrix}.$$

$$(4.14)$$

It remains only to premultiply M by $K^{-1}(V_L)$ and postmultiply by $K(V_R)$. We consider two cases separately.

If the asymptotic potentials V_L, V_R are both less than the energy E of the particle, the matrices K are complex:

$$T = \frac{1}{2} \begin{bmatrix} 1 & \frac{1}{ik_L} \\ 1 & \frac{1}{-ik_L} \end{bmatrix} \begin{bmatrix} m_{11} & m_{12} \\ m_{21} & m_{22} \end{bmatrix} \begin{bmatrix} 1 & 1 \\ ik_R & -ik_R \end{bmatrix} \tag{4.15}$$

$$= \frac{1}{2} \begin{bmatrix} m_{11} + \frac{k_R}{k_L}m_{22} + ik_R m_{12} + \frac{m_{21}}{ik_L} & m_{11} - \frac{k_R}{k_L}m_{22} - ik_R m_{12} + \frac{m_{21}}{ik_L} \\ m_{11} - \frac{k_R}{k_L}m_{22} + ik_R m_{12} - \frac{m_{21}}{ik_L} & m_{11} + \frac{k_R}{k_L}m_{22} - ik_R m_{12} - \frac{m_{21}}{ik_L} \end{bmatrix} . \tag{4.16}$$

This matrix with complex matrix elements can be written in the simpler-looking form

$$T(E) = \begin{bmatrix} \alpha & \beta \\ \bar{\beta} & \bar{\alpha} \end{bmatrix} , \tag{4.17}$$

where $\bar{\alpha}$ is the complex conjugate of the complex number α, and similarly for β. These two complex numbers are explicitly given by $\alpha = \alpha_R + i\alpha_I$, $\beta = \beta_R + i\beta_I$:

$$\begin{aligned} 2\alpha_R &= +m_{11} + \frac{k_R}{k_L}m_{22} , \\ 2\beta_R &= +m_{11} - \frac{k_R}{k_L}m_{22} , \\ 2\alpha_I &= +k_R m_{12} - \frac{m_{21}}{k_L} , \\ 2\beta_I &= -k_R m_{12} - \frac{m_{21}}{k_L} . \end{aligned} \tag{4.18}$$

It is a simple matter to verify that

$$|\alpha|^2 - |\beta|^2 = \frac{k_R}{k_L} \tag{4.19}$$

If the asymptotic potentials V_L, V_R are both greater than the energy E of the particle, then the matrices K are real and

$$T = \frac{1}{2} \begin{bmatrix} 1 & \frac{-1}{\kappa_L} \\ 1 & \frac{1}{\kappa_L} \end{bmatrix} \begin{bmatrix} m_{11} & m_{12} \\ m_{21} & m_{22} \end{bmatrix} \begin{bmatrix} 1 & 1 \\ -\kappa_R & \kappa_R \end{bmatrix}$$

$$= \frac{1}{2} \begin{bmatrix} m_{11} + \frac{\kappa_R}{\kappa_L}m_{22} - \kappa_R m_{12} - \frac{m_{21}}{\kappa_L} & m_{11} - \frac{\kappa_R}{\kappa_L}m_{22} + \kappa_R m_{12} - \frac{m_{21}}{\kappa_L} \\ m_{11} - \frac{\kappa_R}{\kappa_L}m_{22} - \kappa_R m_{12} + \frac{m_{21}}{\kappa_L} & m_{11} + \frac{\kappa_R}{\kappa_L}m_{22} + \kappa_R m_{12} + \frac{m_{21}}{\kappa_L} \end{bmatrix}$$

$$= \begin{bmatrix} \alpha_1 + \alpha_2 & \beta_1 + \beta_2 \\ \beta_1 - \beta_2 & \alpha_1 - \alpha_2 \end{bmatrix} , \tag{4.20}$$

where the real numbers $\alpha_1, \alpha_2, \beta_1, \beta_1$ are given by

$$2\alpha_1 = +m_{11} + \frac{\kappa_R}{\kappa_L}m_{22},$$

$$2\beta_1 = +m_{11} - \frac{\kappa_R}{\kappa_L}m_{22},$$

$$2\alpha_2 = -\kappa_R m_{12} - \frac{m_{21}}{\kappa_L},$$

$$2\beta_2 = +\kappa_R m_{12} - \frac{m_{21}}{\kappa_L}. \tag{4.21}$$

It is a simple matter to verify that

$$\begin{aligned}
\det T &= (\alpha_1^2 - \alpha_2^2) - (\beta_1^2 - \beta_2^2) \\
&= (\alpha_1^2 + \beta_2^2) - (\alpha_2^2 + \beta_1^2) \\
&= \frac{\kappa_R}{\kappa_L}.
\end{aligned} \tag{4.22}$$

The transfer matrices (4.16) and (4.20) are related by

$$\begin{array}{ccc}
(4.16) & \leftrightarrow & (4.20) \\
+ik & \leftrightarrow & -\kappa
\end{array} \tag{4.23}$$

It happens frequently that the determinant of the transfer matrix must be computed. This is a relatively simple task, as the determinant of a product of matrices is the product of the determinants of the individual matrices, and for a nonsingular matrix M, $\det M^{-1} = 1/(\det M)$. To compute $\det T_{0,N+1}$ from (4.8) we observe that the determinants of all matrices $E(V; a)$ are 1 (i.e., Table 2.1). In addition, for every matrix $K(V_j)$ there is a matrix $K^{-1}(V_j)$, except on the far left and the far right, where $K^{-1}(V_L)$ and $K(V_R)$ are unmatched. Therefore,

$$\det T_{0,N+1} = \frac{\det K(V_R)}{\det K(V_L)}. \tag{4.24}$$

This result can be seen even more easily from (4.12) or (4.13), since $\det M_j = 1$ (see Table 4.1).

4.3   SCATTERING MATRICES

The complex numbers A_L, B_L are probability amplitudes for particles moving toward and away from the scattering potential on the left; B_R and A_R are probability amplitudes for a particle moving toward and away from the scattering potential on the right. These four numbers are not independent: there are two linear relations among them. These are provided by the transfer matrix T, which relates the pair $\begin{bmatrix} A_L \\ B_L \end{bmatrix}$ on the left with the pair $\begin{bmatrix} A_R \\ B_R \end{bmatrix}$ on the right.

There is another useful relation among these four amplitudes. This relates the amplitudes $\begin{bmatrix} A_L \\ B_R \end{bmatrix}$ for particles moving toward the scattering potential to the amplitudes $\begin{bmatrix} A_R \\ B_L \end{bmatrix}$ for particles leaving the scattering region. This linear relation defines the scattering matrix, or S-matrix $S(E)$:

$$\begin{bmatrix} A_R \\ B_L \end{bmatrix} = S \begin{bmatrix} A_L \\ B_R \end{bmatrix} = \begin{bmatrix} S_{11}(E) & S_{12}(E) \\ S_{21}(E) & S_{22}(E) \end{bmatrix} \begin{bmatrix} A_L \\ B_R \end{bmatrix}. \qquad (4.25)$$

The T- and S-matrices have dual interpretations. The transfer matrix relates amplitudes in space—on the left and on the right of the scattering region. The scattering matrix relates amplitudes before the interaction with those after the interaction. This duality is illustrated in Fig. 4.3.

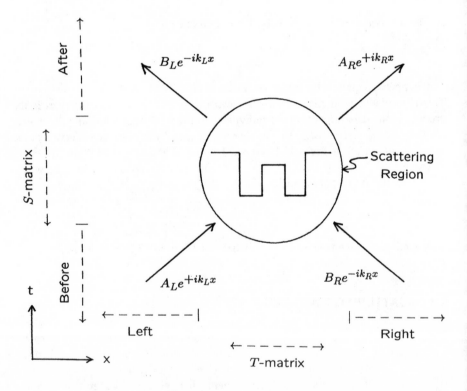

Fig. 4.3 The transfer matrix relates amplitudes on the left of the scattering region with those on the right. The scattering matrix relates incoming amplitudes ("before") with outgoing amplitudes ("after").

It is a simple matter to construct the matrix elements of the S-matrix from those of the T-matrix. We first write out the two equations summarized by the T-matrix:

$$
\begin{aligned}
A_L &= t_{11}A_R + t_{12}B_R, \\
B_L &= t_{21}A_R + t_{22}B_R.
\end{aligned}
\tag{4.26}
$$

Then we regroup the complex amplitudes, placing the amplitudes $\begin{bmatrix} A_R \\ B_L \end{bmatrix}$ for out-going waves on the left and the incoming amplitudes $\begin{bmatrix} A_L \\ B_R \end{bmatrix}$ on the right:

$$
\begin{aligned}
-t_{11}A_R &= -A_L + t_{12}B_R \\
-t_{21}A_R + B_L &= t_{22}B_R,
\end{aligned}
\tag{4.27}
$$

$$
\begin{bmatrix} -t_{11} & 0 \\ -t_{21} & 1 \end{bmatrix} \begin{bmatrix} A_R \\ B_L \end{bmatrix} = \begin{bmatrix} -1 & t_{12} \\ 0 & t_{22} \end{bmatrix} \begin{bmatrix} A_L \\ B_R \end{bmatrix}.
\tag{4.28}
$$

The linear relation we desire is obtained by multiplying by the inverse of the matrix on the left:

$$
\begin{bmatrix} A_R \\ B_L \end{bmatrix} = \begin{bmatrix} -t_{11} & 0 \\ -t_{21} & 1 \end{bmatrix}^{-1} \begin{bmatrix} -1 & t_{12} \\ 0 & t_{22} \end{bmatrix} \begin{bmatrix} A_L \\ B_R \end{bmatrix}.
\tag{4.29}
$$

The result is

$$
S = \begin{bmatrix} \dfrac{1}{t_{11}} & -\dfrac{t_{12}}{t_{11}} \\[2ex] \dfrac{t_{21}}{t_{11}} & \dfrac{\det(T)}{t_{11}} \end{bmatrix} = \begin{bmatrix} s_{11}(E) & s_{12}(E) \\[1ex] s_{21}(E) & s_{22}(E) \end{bmatrix}.
\tag{4.30}
$$

Conservation of momentum provides the following quantities conserved by the T- and S-matrices:

$$
\begin{array}{ll}
T & k_L|A_L|^2 - k_L|B_L|^2 = k_R|A_R|^2 - k_R|B_R|^2, \\
S & k_L|A_L|^2 + k_R|B_R|^2 = k_R|A_R|^2 + k_L|B_L|^2.
\end{array}
\tag{4.31}
$$

When the asymptotic potentials on the left and right are equal, $V_L = V_R$, these conservation laws simplify to

$$
\begin{array}{ll}
T & |A_L|^2 - |B_L|^2 = |A_R|^2 - |B_R|^2 \\
S & |A_L|^2 + |B_R|^2 = |A_R|^2 + |B_L|^2 = 1.
\end{array}
\tag{4.32}
$$

One additional linear relation among two pairs of amplitudes is possible. This relates the amplitudes for right-going waves with the amplitudes for left-going waves:

$$
\begin{bmatrix} A_L \\ A_R \end{bmatrix} = U \begin{bmatrix} B_R \\ B_L \end{bmatrix}.
\tag{4.33}
$$

This last relation is almost never used.

In this work we will deal entirely with transfer matrices. However, there are many one-dimensional quantum mechanical problems that are not elementary and that can only be treated with S-matrices.

5

Momentum Conservation

We have seen in chapter 2 that the wavefunction $\Phi_1(x) = e^{+ikx}$ represents a particle traveling to the right in a region of constant potential $V < E$ with a momentum $p = +\hbar k, k = \sqrt{2m(E - V)/\hbar^2}$. Similarly, $\Phi_2(x) = e^{-ikx}$ represents a particle traveling to the left, with a momentum $p = -\hbar k$.

The most general wavefunction in such a region is a complex linear superposition of the two particular solutions,

$$\Phi(x) = Ae^{+ikx} + Be^{-ikx} . \tag{5.1}$$

The complex number A is the probability amplitude for finding the particle moving to the right with momentum $\hbar k$. Its absolute square, $|A|^2 = \overline{A}A = A^*A$, is the probability for finding the particle with momentum $+\hbar k$. Similarly, B is the probability amplitude for finding the particle with momentum $-\hbar k$, and $|B|^2 = \overline{B}B = B^*B$ is the probability for finding the particle with momentum $-\hbar k$.

Since a measurement of particle momentum will yield only one of the two results $p = +\hbar k$ or $p = -\hbar k$, the probability of finding the particle of energy $E = (\hbar k)^2/2m + V$ with momentum either $+\hbar k$ or $-\hbar k$ is one. Therefore

$$|A|^2 + |B|^2 = 1 . \tag{5.2}$$

The average particle momentum in the region of constant potential V is the momentum $\hbar k$ multiplied by the probability that the particle has momentum $\hbar k$, plus the momentum $-\hbar k$ multiplied by the probability that the particle has momentum $-\hbar k$:

$$p_{av} = <\hat{p}> = (+\hbar k)\text{Pr}(\hbar k) + (-\hbar k)\text{Pr}(-\hbar k) = \hbar k(|A|^2 - |B|^2) . \tag{5.3}$$

We will now describe the quantum mechanical version of the law of momentum conservation. We start with a simple problem (Fig. 5.1). In the left-hand region with constant potential $V_1 < E$ the wavefunction and average momentum are

$$\begin{aligned} \Phi(x) &= Ae^{+ik_1 x} + Be^{-ik_1 x} , \\ <\hat{p}> &= \hbar k_1 (|A|^2 - |B|^2) . \end{aligned} \qquad (5.4)$$

In the right-hand region with constant potential $V_2 < E$ we have

$$\begin{aligned} \Phi(x) &= Ce^{+ik_2 x} + De^{-ik_2 x} , \\ <\hat{p}> &= \hbar k_2 (|C|^2 - |D|^2) . \end{aligned} \qquad (5.5)$$

We want to show that the average momentum in the left-hand region is equal to the average momentum in the right-hand region:

$$\hbar k_1 (|A|^2 - |B|^2) = \hbar k_2 (|C|^2 - |D|^2) . \qquad (5.6)$$

The verification of (5.6) is easily carried out using transfer matrix methods. The complex amplitudes (A, B) and (C, D) are related by

$$\begin{bmatrix} A \\ B \end{bmatrix} = T_{12} \begin{bmatrix} C \\ D \end{bmatrix} = E^{-1}(V_1; a) K^{-1}(V_1) K(V_2) E(V_2; a) \begin{bmatrix} C \\ D \end{bmatrix} . \qquad (5.7)$$

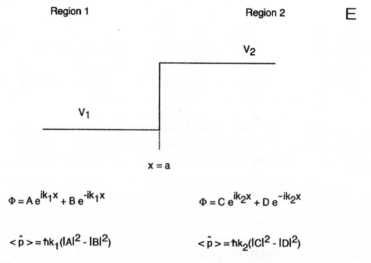

Fig. 5.1 The mean value of the particle momentum in region 1 is the same as the mean value of the momentum in region 2.

We could simplify this calculation by choosing the break point a to be at the origin of coordinates. However, it will be useful to allow a to be nonzero. Then the complex phase factors in the E matrices can be absorbed into the amplitudes (A, B) and (C, D) as follows:

$$\begin{bmatrix} A' \\ B' \end{bmatrix} = E(V_1; a) \begin{bmatrix} A \\ B \end{bmatrix} = \begin{pmatrix} e^{+ik_1 a} A \\ e^{-ik_1 a} B \end{pmatrix},$$

$$\begin{bmatrix} C' \\ D' \end{bmatrix} = E(V_2; a) \begin{bmatrix} C \\ D \end{bmatrix} = \begin{pmatrix} e^{+ik_2 a} C \\ e^{-ik_2 a} D \end{pmatrix}. \tag{5.8}$$

These phase factors will be unimportant in the final analysis since (5.6) involves only the absolute squares of the complex amplitudes.

The result (5.7) then reduces to

$$\begin{bmatrix} A' \\ B' \end{bmatrix} = \frac{1}{2} \begin{pmatrix} 1 & \frac{1}{ik_1} \\ 1 & \frac{1}{-ik_1} \end{pmatrix} \begin{pmatrix} 1 & 1 \\ ik_2 & -ik_2 \end{pmatrix} \begin{bmatrix} C' \\ D' \end{bmatrix}$$

$$= \begin{bmatrix} \frac{k_1+k_2}{2k_1} & \frac{k_1-k_2}{2k_1} \\ \frac{k_1-k_2}{2k_1} & \frac{k_1+k_2}{2k_1} \end{bmatrix} \begin{bmatrix} C' \\ D' \end{bmatrix}. \tag{5.9}$$

Now we compute $|A'|^2, |B'|^2$, and take their difference

$$|A'|^2 = \left(\frac{k_1 + k_2}{2k_1} \right)^2 |C'|^2 + \left(\frac{k_1 - k_2}{2k_1} \right)^2 |D'|^2 + \frac{k_1^2 - k_2^2}{(2k_1)^2} (\overline{C'}D' + C'\overline{D'})$$

$$|B'|^2 = \left(\frac{k_1 - k_2}{2k_1} \right)^2 |C'|^2 + \left(\frac{k_1 + k_2}{2k_1} \right)^2 |D'|^2 + \frac{k_1^2 - k_2^2}{(2k_1)^2} (\overline{C'}D' + C'\overline{D'})$$

$$|A'|^2 - |B'|^2 = \frac{k_2}{k_1} (|C'|^2 - |D'|^2). \tag{5.10}$$

This last equation is what we have set out to prove when we recognize that $|A'|^2 = |A|^2$, and so on.

We now prove momentum conservation in the general case where $E > V_L$, $E > V_R$, shown in Fig. 5.2. The potential may be approximated by a piecewise constant potential by choosing the breakpoints close enough ($a_{j+1} - a_j \sim \epsilon, \epsilon$ very small) and allowing N to be large enough. The transfer matrix $T_{0,N+1}$ can be expressed in the form (4.12). The phase factors in the exponentials that occur at the ends of $T_{0,N+1}$ may be absorbed into the amplitudes, as in (5.8). The product of the matrices M_j that occur in the interior of (4.12) need not be computed explicitly. We have only to observe that each M_j is real and has determinant $+1$. The product of real matrices with determinant $+1$ is itself a real matrix with determinant $+1$. Thus, no matter what the potential,

$$M_1 M_2 \ldots M_N = \begin{bmatrix} a & b \\ c & d \end{bmatrix}, \quad ad - bc = 1, \quad a, b, c, d \quad \text{real}. \tag{5.11}$$

The relation between (A_0, B_0) and (A_{N+1}, B_{N+1}) or (A_L, B_L) and (A_R, B_R) is therefore

$$\begin{bmatrix} A'_L \\ B'_L \end{bmatrix} = \frac{1}{2} \begin{pmatrix} 1 & \frac{1}{ik_L} \\ 1 & \frac{1}{-ik_L} \end{pmatrix} \begin{bmatrix} a & b \\ c & d \end{bmatrix} \begin{bmatrix} 1 & 1 \\ ik_R & -ik_R \end{bmatrix} \begin{bmatrix} A'_R \\ B'_R \end{bmatrix}$$

$$= \begin{bmatrix} \alpha & \beta \\ \overline{\beta} & \overline{\alpha} \end{bmatrix} \begin{bmatrix} A'_R \\ B'_R \end{bmatrix}, \tag{5.12}$$

$$2\alpha = \left(a + d\frac{k_R}{k_L}\right) + i\left(+bk_R - \frac{c}{k_L}\right),$$

$$2\beta = \left(a - d\frac{k_R}{k_L}\right) + i\left(-bk_R - \frac{c}{k_L}\right). \tag{5.13}$$

Again we compute $|A'_L|^2, |B'_L|^2$ and take their difference

$$|A'_L|^2 = |\alpha|^2|A'_R|^2 + |\beta|^2|B'_R|^2 + \overline{\alpha}\beta\,\overline{A'_R}B'_R + \alpha\overline{\beta}A'_R\overline{B'_R},$$

$$|B'_L|^2 = |\beta|^2|A'_R|^2 + |\alpha|^2|B'_R|^2 + \overline{\alpha}\beta\overline{A'_R}B'_R + \alpha\overline{\beta}A'_R\overline{B'_R},$$

$$|A'_L|^2 - |B'_L|^2 = (|\alpha|^2 - |\beta|^2)(|A'_R|^2 - |B'_R|^2)$$

$$\stackrel{\text{by (5.12)}}{=} (ad - bc)\frac{k_R}{k_L}(|A'_R|^2 - |B'_R|^2)$$

$$\stackrel{\text{by (5.11)}}{=} \frac{k_R}{k_L}(|A'_R|^2 - |B'_R|^2). \tag{5.14}$$

Since $|A'_L|^2 = |A_L|^2$ and so on, we have the desired result that (average) momentum is conserved on transmission through a barrier of arbitrary shape

$$\hbar k_L(|A_L|^2 - |B_L|^2) = \hbar k_R(|A_R|^2 - |B_R|^2). \tag{5.15}$$

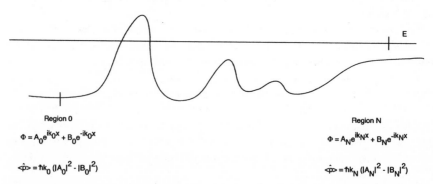

Region 0

$\Phi = A_0 e^{ik_0 x} + B_0 e^{-ik_0 x}$

$\langle \dot{p} \rangle = \hbar k_0 \left(|A_0|^2 - |B_0|^2\right)$

Region N

$\Phi = A_N e^{ik_N x} + B_N e^{-ik_N x}$

$\langle \dot{p} \rangle = \hbar k_N \left(|A_N|^2 - |B_N|^2\right)$

Fig. 5.2 In the general scattering case, the average momentum of the particle in the asymptotic left-hand region is the same as the average momentum in the asymptotic right-hand region, no matter what the shape of the potential.

Preview of Boundary Conditions

We will solve the time-independent Schrödinger equation in one dimension with piecewise constant potentials subject to three distinct boundary conditions:

Part II   Scattering
Part III   Bound States
Part IV   Periodic Potentials

Each boundary condition imposes a different condition on the transfer matrix.

6.1 BOUNDARY CONDITION FOR SCATTERING

In the case of scattering (Fig. 6.1) the wavefunctions in the asymptotic left-hand and right-hand regions are

$$\Phi_L(x) = A_L e^{+ik_L x} + B_L e^{-ik_L x} , \quad \Phi_R(x) = A_R e^{+ik_R x} + B_R e^{-ik_R x} . \quad (6.1)$$

The amplitudes for the wavefunction on the left- and right-hand side of the potential are related by the transfer matrix

$$\begin{bmatrix} A_L \\ B_L \end{bmatrix} = T \begin{bmatrix} A_R \\ B_R \end{bmatrix} = \begin{bmatrix} t_{11}(E) & t_{12}(E) \\ t_{21}(E) & t_{22}(E) \end{bmatrix} \begin{bmatrix} A_R \\ B_R \end{bmatrix} . \quad (6.2)$$

We assume that the constant potentials on the left and right of the scattering potential are equal. We also assume that a particle is incident from the left with nonzero probability amplitude ($A_L \neq 0$), but not from the right ($B_R = 0$). There is some probability amplitude (A_R) that the particle is transmitted through the barrier, and some amplitude (B_L) that it is reflected. This provides a simple relation between the amplitudes

$$\begin{bmatrix} A_L \\ B_L \end{bmatrix} = \begin{bmatrix} t_{11} & t_{12} \\ t_{21} & t_{22} \end{bmatrix} \begin{bmatrix} A_R \\ 0 \end{bmatrix} \Longrightarrow \begin{matrix} A_L & = & t_{11}(E)A_R \\ B_L & = & t_{21}(E)A_R \end{matrix} \quad . \tag{6.3}$$

The squares of A_R, B_L describe the transmission probability $T(E)$ and reflection probability $R(E)$ for the particle incident on the scattering potential

$$\begin{matrix} T(E) & = & |A_R/A_L|^2 & = & 1/|t_{11}(E)|^2 \\ R(E) & = & |B_L/A_L|^2 & = & |t_{21}(E)|^2/|t_{11}(E)|^2 \end{matrix} . \tag{6.4}$$

We remark that these results can be determined from the S-matrix (4.25) with (4.30).

6.2   BOUNDARY CONDITION FOR BOUND STATES

In the case of bound states (Fig. 6.2) the wavefunctions in the asymptotic left- and right-hand regions, forbidden to a classical particle, are

$$\Phi_L(x) = A_L e^{-\kappa_L x} + B_L e^{+\kappa_L x} , \quad \Phi_R(x) = A_R e^{-\kappa_R x} + B_R e^{+\kappa_R x} . \tag{6.5}$$

In order to have a wavefunction that is bounded by the classically forbidden right-hand region, $B_R = 0$. Then

$$\begin{bmatrix} A_L \\ B_L \end{bmatrix} = \begin{bmatrix} t_{11} & t_{12} \\ t_{21} & t_{22} \end{bmatrix} \begin{bmatrix} A_R \\ 0 \end{bmatrix} \Longrightarrow \begin{matrix} A_L & = & t_{11}(E)A_R \\ B_L & = & t_{21}(E)A_R \end{matrix} \quad . \tag{6.6}$$

In order to have a wavefunction that is bounded by the classically forbidden left-hand region, $A_L = 0$. Since $A_R \neq 0$ (otherwise the wavefunction would vanish everywhere), $t_{11}(E)$ must be zero. Thus, the zeros of $t_{11}(E)$ define the energies at which the potential supports bound states. This result can also be determined from the S-matrix (4.25) with (4.30).

6.3   BOUNDARY CONDITION FOR PERIODIC POTENTIALS

Many solids are adequately approximated by a long sequence of identical potentials (Fig. 6.3). If the transfer matrix for each unit cell in this potential is $T(E)$, then the transfer matrix for N identical cells "in series" is $[T(E)]^N$.

For reasons that will be justified later, it is useful to assume periodic boundary conditions. That is, we identify the wavefunction at one end of the potential with that at the other, so that $\begin{bmatrix} A \\ B \end{bmatrix}_0 = \begin{bmatrix} A \\ B \end{bmatrix}_N$:

$$\begin{bmatrix} A \\ B \end{bmatrix}_0 = [T(E)]^N \begin{bmatrix} A \\ B \end{bmatrix}_N . \tag{6.7}$$

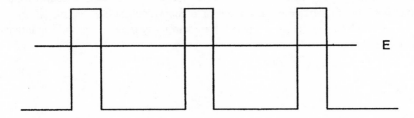

Fig. 6.1 Typical boundary conditions for scattering

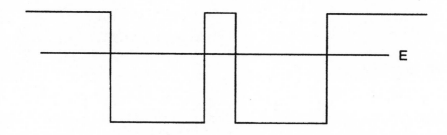

Fig. 6.2 A typical arrangement for bound states

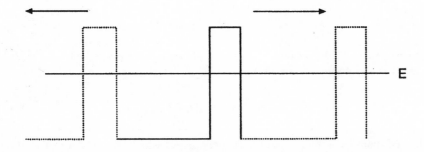

Fig. 6.3 A typical periodic potential geometry

This identification requires $[T(E)]^N = I_2$, the unit 2×2 matrix. Therefore, the problem of identifying the allowed states and energies for a periodic potential reduces to the problem of determining the values of E for which $[T(E)]^N = I_2$.

To approach this problem, we search for a similarity transformation S that diagonalizes $T(E)$:

$$S\,[T(E)]^N\,S^{-1} \;=\; SI_2S^{-1}\,,$$

$$\left[ST(E)S^{-1}\right]^N \;=\; I_2\,, \tag{6.8}$$

$$\begin{bmatrix} \lambda_1 & 0 \\ 0 & \lambda_2 \end{bmatrix}^N \;=\; \begin{bmatrix} 1 & 0 \\ 0 & 1 \end{bmatrix}.$$

Therefore, the condition simplifies to $\lambda_j^N = 1$ ($j = 1, 2$). Since the transfer matrix $T(E)$ is unimodular, the product of the two eigenvalues is $+1$, so that $\lambda_1 = \lambda = \lambda_2^{-1}$. The eigenvalues of the unit cell transfer matrix $T(E)$ are determined from

$$\det \left| \begin{bmatrix} t_{11} & t_{12} \\ t_{21} & t_{22} \end{bmatrix} - \lambda \begin{bmatrix} 1 & 0 \\ 0 & 1 \end{bmatrix} \right| \;=\; \lambda^2 - (t_{11} + t_{22})\lambda + (t_{11}t_{22} - t_{12}t_{21})$$

$$= \lambda^2 - \lambda \operatorname{tr}(T) + \det(T)\,. \tag{6.9}$$

Since $\det(T) = 1$,

$$\begin{aligned} \lambda &= \tfrac{1}{2}\operatorname{tr}(T) \pm \sqrt{\left(\tfrac{1}{2}\operatorname{tr}(T)\right)^2 - 1} \\ &= \tfrac{1}{2}\operatorname{tr}(T) \pm i\sqrt{1 - \left(\tfrac{1}{2}\operatorname{tr}(T)\right)^2}\,. \end{aligned} \tag{6.10}$$

If we define an angle ϕ by

$$\cos\phi \;=\; \tfrac{1}{2}\operatorname{tr}(T)\,,$$

$$\sin\phi \;=\; \sqrt{1 - \left(\tfrac{1}{2}\operatorname{tr}(T)\right)^2}\,, \tag{6.11}$$

$$\lambda^{\pm 1} \;=\; \cos\phi \pm i\sin\phi \;=\; e^{\pm i\phi}\,.$$

It is easily seen that the periodic boundary condition $\lambda^N = 1$ is satisfied if $N\phi = 2\pi k$, where k is an integer. The result is

$$\operatorname{tr}(T) = t_{11} + t_{22} = 2\cos\left(2\pi\frac{k}{N}\right)\,. \tag{6.12}$$

Allowed states exist for values of the energy, E, for which the transfer matrix of the unit cell, $T(E)$, satisfies the condition (6.12).

In the following parts of this book we will explore the implications of these three types of boundary conditions.

7

Units

Most of the calculations that will be carried out in Parts II, III, and IV of this work will be numerical computations based on equations (6.4), (6.6), and (6.12). In order to compute the transfer matrices, and in particular the real 2×2 matrices $M(V;\delta)$ for each region of a piecewise constant potential, we must provide information about the energy and width of each piece of the potential. This means we must adopt a system of units for measuring energy and length.

For an electron with mass $m = 0.911 \times 10^{-27}$ gm and charge $q = -1.602 \times 10^{-19}$ Coul, a useful unit in which to measure energy is the electron-volt (eV). This is the energy gained (or lost) by an electron when it moves through a potential difference of one volt:

$$
\begin{aligned}
\Delta E &= |qV| \\
&= |(-1.602 \times 10^{-19}\text{ Coul}) \times (1\text{ volt})| \\
&= 1.602 \times 10^{-19}\text{ J } (= \text{kg m}^2/\text{sec}^2) \\
&= 1.602 \times 10^{-12}\text{ erg } (= \text{gm cm}^2/\text{sec}^2)\,.
\end{aligned}
\tag{7.1}
$$

Electron-volts are convenient units because it is very easy to change the energy of a charged particle by changing the imposed potential difference (voltage) across the region through which the charge moves.

To determine a useful length scale, we search for a length a for which the dimensionless product ka is approximately 1 for an electron moving with an energy of 1 eV:

$$
\begin{aligned}
\frac{\hbar^2 k^2}{2m} &= E = 1\,\text{eV}\,, \\
k^2 &= \frac{2mE}{\hbar^2} \\
&= \frac{2(0.911 \times 10^{-27}\,\text{gm})(1.602 \times 10^{-12}\,\text{erg})}{(1.054 \times 10^{-27}\,\text{erg sec})^2} \\
&= 2.626 \times 10^{+15}\,\text{cm}^{-2}\,, \\
k &= 0.512 \times 10^{+8}\,\text{cm}^{-1}\,.
\end{aligned}
\tag{7.2}
$$

Therefore a useful length scale is 10^{-8} cm = 1 Å (angstrom). To give some perspective to these units, the "diameter" of a hydrogen atom in its ground state is about 1 Å and the electron is bound to the proton in this state with an energy of about 13.6 eV.

All energies will be measured in electron-volts, and all lengths will be measured in angstroms throughout the remainder of this work. In these units the relation between k and E is

$$
k = \sqrt{\frac{2m}{\hbar^2} E} = \sqrt{\frac{2m}{\hbar^2}\, q\, \frac{E}{q}}\,.
\tag{7.3}
$$

In this expression, E/q is measured in terms of electron-volts. Using the physical values given above,

$$
k = 0.5125\sqrt{E}\,,
\tag{7.4}
$$

with E measured in electron-volts.

Part II

Scattering

8

Boundary Conditions

In many systems of interest, particles are incident on a target from some particular direction. For example, in a scattering experiment electrons might be beamed from an electron gun at a hydrogen target. Some of the electrons may propagate through the target, others may be reflected by the target. In this chapter we learn how to impose such physically realistic boundary conditions on one-dimensional problems.

We begin by representing the target by a stationary one-dimensional potential with asymptotic values V_L, V_R on the left and right. The particle incident from the left is represented by the wavefunction $A_L e^{+ik_L x}$, $k_L = [2m(E - V_L)/\hbar^2]^{1/2}$. The wavefunction representing the reflected particle is $B_L e^{-ik_L x}$; the wavefunction representing the transmitted particle is $A_R e^{+ik_R x}$. Thus, the boundary condition for a particle incident on the target from the left is $B_R = 0$.

We have seen in chapter 5 that the law of momentum conservation can be written as

$$\hbar k_L(|A_L|^2 - |B_L|^2) = \hbar k_R(|A_R|^2 - |B_R|^2) . \tag{8.1}$$

For a particle incident from the left ($B_R = 0$), this expression can be written

$$\left|\frac{B_L}{A_L}\right|^2 + \frac{k_R}{k_L}\left|\frac{A_R}{A_L}\right|^2 = 1 . \tag{8.2}$$

Since B_L is the probability amplitude for the reflected particle and A_L is the probability amplitude for the incident particle, $|B_L/A_L|^2$ has a natural interpretation as the probability that the incident particle is reflected. If the particle is not reflected or absorbed, it is transmitted. Therefore, $(k_R/k_L)|A_R/A_L|^2$ must be interpreted as the probability that the incident particle is transmitted. In summary, the transmission (T) and reflection (R) probabilities are

$$T = \frac{k_R}{k_L}\left|\frac{A_R}{A_L}\right|^2, \qquad R = \left|\frac{B_L}{A_L}\right|^2, \qquad T + R = 1. \qquad (8.3)$$

Since

$$\begin{bmatrix} A_L \\ B_L \end{bmatrix} = \begin{bmatrix} t_{11}(E) & t_{12}(E) \\ t_{21}(E) & t_{22}(E) \end{bmatrix}\begin{bmatrix} A_R \\ B_R = 0 \end{bmatrix} = \begin{bmatrix} t_{11}(E)A_R \\ t_{21}(E)A_R \end{bmatrix}, \qquad (8.4)$$

it follows that

$$T = \frac{k_R}{k_L}\left|\frac{1}{t_{11}(E)}\right|^2. \qquad (8.5)$$

This condition can be made more explicit by expressing $T(E)$ in terms of the product of matrices $M(V_j; \delta_j)$ for the interior pieces of the potential, and the pair of matrices for the two asymptotic regions

$$\begin{bmatrix} A \\ B \end{bmatrix}_L = \begin{bmatrix} e^{+ik_L a_L} & 0 \\ 0 & e^{-ik_L a_L} \end{bmatrix}^{-1}\begin{bmatrix} 1 & 1 \\ +ik_L & -ik_L \end{bmatrix}^{-1}\prod_{j=1}^{N} M(V_j; \delta_j)$$

$$\times \begin{bmatrix} 1 & 1 \\ +ik_R & -ik_R \end{bmatrix}\begin{bmatrix} e^{+ik_R a_R} & 0 \\ 0 & e^{-ik_R a_R} \end{bmatrix}\begin{bmatrix} A \\ B \end{bmatrix}_R. \qquad (8.6)$$

It is convenient to absorb the exponentials into the definition of the amplitudes. Carrying out the remaining matrix multiplications, we obtain

$$\begin{bmatrix} A \\ B \end{bmatrix}_L' = \begin{bmatrix} \alpha_R + i\alpha_I & \beta_R + i\beta_I \\ \beta_R - i\beta_I & \alpha_R - i\alpha_I \end{bmatrix}\begin{bmatrix} A \\ B \end{bmatrix}_R', \qquad (8.7)$$

$$2\alpha_R = +m_{11} + \frac{k_R}{k_L}m_{22}, \qquad 2\alpha_I = +k_R m_{12} - \frac{m_{21}}{k_L},$$

$$2\beta_R = +m_{11} - \frac{k_R}{k_L}m_{22}, \qquad 2\beta_I = -k_R m_{12} - \frac{m_{21}}{k_L},$$

where $A_L' = A_L e^{+ik_L a_L}$ and so on, and a_L, a_R are the left- and right-hand boundaries of the potential. Then $|A_L'|^2 = |A_L|^2$ and so on, so that $T(E) = (k_R/k_L)|A_R'/A_L'|^2$. As a result, the transmission probability is given by

$$T(E) = \frac{4(k_R/k_L)}{[m_{11} + (k_R/k_L)m_{22}]^2 + [k_R m_{12} - m_{21}/k_L]^2}. \qquad (8.8)$$

In the typically encountered case in which $V_L = V_R$, $k_L = k_R = k$ and this expression simplifies to

$$T(E) = \frac{4}{(m_{11} + m_{22})^2 + (km_{12} - m_{21}/k)^2}. \qquad (8.9)$$

This is the expression we use to compute almost all transmission probabilities in this work.

9

A Simple Example

To illustrate these results, we compute the transmission probability for the rectangular barrier with constant potential V, width δ, shown in the inset of Fig. 9.1. We assume $V_L = V_R = 0$. The transfer matrix is given by (4.12). Since $k_L = k_R = k = \sqrt{2mE/\hbar^2}$, the transmission probability is $T = |1/t_{11}(E)|^2$. The matrix elements $m_{ij}(E)$ for the single intermediate potential can be seen by inspection from Table 4.1. There are three cases to consider:

$E < V$: $\kappa = \sqrt{2m(V - E)/\hbar^2}$,

$$t_{11}(E) = \cosh \kappa\delta - \frac{i}{2}\left(-\frac{\kappa}{k} + \frac{k}{\kappa}\right)\sinh \kappa\delta , \qquad (9.1)$$

$$T(E) = \frac{1}{\cosh^2 \kappa\delta + \frac{1}{4}\left(\frac{\kappa}{k} - \frac{k}{\kappa}\right)^2 \sinh^2 \kappa\delta}$$

$$= \frac{1}{1 + \frac{1}{4}\left(\frac{\kappa}{k} + \frac{k}{\kappa}\right)^2 \sinh^2 \kappa\delta} ; \qquad (9.2)$$

$E = V$:

$$t_{11}(E) = 1 - ik\delta/2 , \qquad (9.3)$$

$$T(E) = \frac{1}{1 + (k\delta/2)^2} ; \qquad (9.4)$$

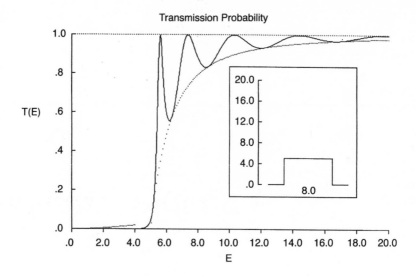

Fig. 9.1 Transmission probability, $T(E)$, as a function of energy for an incident electron of energy E on the barrier shown in the inset. The barrier has height 5 eV, width 8 Å. The transmission probability is computed using the analytic expressions (9.1)–(9.6). Shown also are the asymptotic estimate for the tunneling probability given in (9.9) (dotted, to 4 eV) and the lower bound on the transmission probability given by the expression (9.12), (dotted, $E > V$).

$$E > V:\ k' = \sqrt{2m(E - V)/\hbar^2},$$

$$t_{11}(E) \quad = \quad \cos k'\delta - \frac{i}{2}\left(\frac{k'}{k} + \frac{k}{k'}\right)\sin k'\delta\,, \tag{9.5}$$

$$T(E) \quad = \quad \frac{1}{\cos^2 k'\delta + \frac{1}{4}\left(\frac{k'}{k} + \frac{k}{k'}\right)^2 \sin^2 k'\delta}$$

$$= \quad \frac{1}{1 + \frac{1}{4}\left(\frac{k'}{k} - \frac{k}{k'}\right)^2 \sin^2 k'\delta}\,. \tag{9.6}$$

The identities $\cosh^2 x - \sinh^2 x = 1$, $\cos^2 x + \sin^2 x = 1$ have been used to construct $T(E)$ from $t_{11}(E)$ in the cases $E < V, E > V$.

The transmission probability $T(E)$ is plotted as a function of E in Fig. 9.1 for the repelling barrier shown in the inset. The behavior of the transmission probability is not exactly intuitive for anyone whose intuition is developed on classical ($=$ nonquantum) mechanics. Classically, a particle with energy $E < V$ will be reflected at the barrier

with 100% probability. When $E > V$ it will be transmitted with 100% probability.[1]
We observe that for $E > V$ the transmission probability is exactly 1 only at isolated
points, which are the zeros of $\sin k'\delta$.

There is always a nonzero probability for transmission through the barrier even
in the classically forbidden regime $E < V$, where the classical particle is reflected.
Transmission through a classically forbidden region is called quantum mechanical
tunneling, or simply tunneling.

The transmission probability is not reduced to zero by making the barrier thicker
or higher, although it may be dramatically reduced. The asymptotic behavior of the
transmission probability through this barrier is simple to discuss. In the classically
forbidden regime $E < V$, $\kappa\delta = \sqrt{2m(V-E)/\hbar^2}\delta$ and

$$\sinh \kappa\delta \;\rightarrow\; \frac{1}{2}e^{\kappa\delta} ,$$

$$(9.7)$$

$$\left(\frac{\kappa}{k} + \frac{k}{\kappa}\right)^2 = \frac{V-E}{E} + 2 + \frac{E}{V-E} = \frac{V^2}{E(V-E)} . \tag{9.8}$$

Using these approximations in the expression for $T(E)$ given in (9.2), and neglecting
"1" compared with the larger term, the asymptotic dependence of the transmission
probability is

$$T(E) \xrightarrow{\kappa\delta \gg 1} 16\frac{E(V-E)}{V^2} e^{-2\kappa\delta} . \tag{9.9}$$

This argument shows that in the classically forbidden regime the transmission probability drops off exponentially, so that

$$\log T(E) \sim -2\kappa\delta = -2\sqrt{2m(V-E)/\hbar^2}\delta . \tag{9.10}$$

In the classically allowed regime $E > V$ the transmission probability (9.6) is
+1 only at isolated energies at which $\sin k'\delta = 0$, or $k'\delta = n\pi$. Since $k' = \sqrt{2m(E-V)/\hbar^2}$, these occur at energies

$$E_n = V + \frac{\hbar^2}{2m}\left(\frac{n\pi}{\delta}\right)^2 \qquad n = 1, 2, \dots . \tag{9.11}$$

However, as the energy increases, the transmission probability gets closer and closer
to +1. In fact, it oscillates between the upper and lower bounds

$$T_{\text{upper bound}} = 1 ,$$

$$T_{\text{lower bound}} = \frac{1}{1 + \frac{1}{4}\left(\frac{k'}{k} - \frac{k}{k'}\right)^2} = 1 - \left(\frac{V}{2E-V}\right)^2 . \tag{9.12}$$

---

[1]This is strictly true only if the barrier height is a sufficiently slowly varying function of position. A
classical bowling ball with $E > V$ could not pass the barrier in the inset of Fig. 9.1 unless the corners
were rounded considerably.

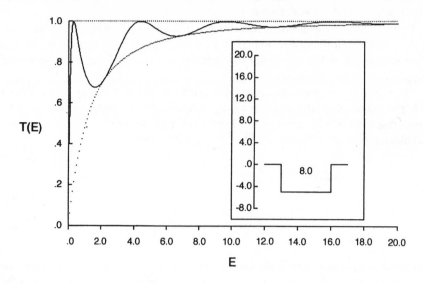

Fig. 9.2 Transmission probability as a function of incident energy for the "attracting barrier" ($V = -5\,\text{eV}$) shown in the inset. Upper and lower bounds on $T(E)$ are shown by dotted lines.

For $E \gg V$ the lower bound approaches the upper bound algebraically (i.e., power law behavior) and the system behaves classically.

It is a simple matter to verify that as E approaches V from below, $T(E)$ given by (9.2) approaches the grazing limit $T(E) = [1 + (k\delta/2)^2]^{-1}$ given by (9.4). Similarly, as E approaches V from above, the limit of $T(E)$ given by (9.6) also approaches this grazing limit.

The transmission probability for an "attracting barrier" has also been computed, and is shown in Fig. 9.2. For this barrier, $V_L = V_R = 0, V < 0$, so the transmission probability is given by

$$
\begin{aligned}
k &= \sqrt{2mE/\hbar^2}\,, \\
k' &= \sqrt{2m(E-V)/\hbar^2} = \sqrt{2m(E+|V|)/\hbar^2}\,, \\
T(E) &= \frac{1}{1 + \frac{1}{4}\left(\frac{k'}{k} - \frac{k}{k'}\right)^2 \sin^2 k'\delta}\,.
\end{aligned}
\tag{9.13}
$$

The bounds on the transmission probability are as given in (9.12).

10

Coding and Validation

Equation (8.3) provides a simple algorithm for computing the transmission and reflection probabilities from the matrix elements of the transfer matrix. However, except for the simplest cases (previous chapter) it is not practical to compute these matrix elements analytically. It is therefore useful to carry out such computations numerically. The algorithm for this numerical computation is straightforward:

- Read in the height V_j and width δ_j of the piecewise constant potential, as well as the asymptotic potential values V_L, V_R on the left and right.

- Choose a value, E, for the incident particle energy $(E > V_L, E > V_R)$.

- Compute the real 2×2 matrix $M(V_j; \delta_j)$ for each piece of the potential.

- Multiply these matrices in the order in which they occur, from left to right: $M(E) = \prod_{j=1}^{j=N} M(V_j; \delta_j)$.

- Compute $t_{11}(E)$ in terms of $m_{ij}(E), k_L, k_R$ using (8.8) or (8.9).

For many purposes it is useful to compute $T(E)$ as a function of the incident particle energy. This means that the computation outlined above must be embedded in a loop that scans over the desired range of values of the incident particle energy.

In Fig. 10.1 we present the numerically computed transmission probability, $T(E)$, for a particle of energy E incident on the barrier shown in the inset. Since this is a very simple barrier, the results are available analytically and have already been plotted in Fig. 9.1. These analytic results are also presented in Fig. 10.1 (dotted curve, slightly offset above the solid curve), for comparison with the numerically computed value of $T(E)$. A similar comparison is made in Fig. 10.2 for the attracting potential treated previously in Fig. 9.2.

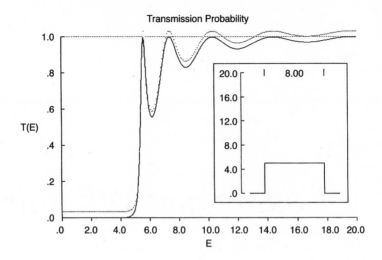

Fig. 10.1 Transmission probability, $T(E)$, computed numerically for an incident electron of energy E, for the potential barrier shown in the inset. This is to be compared with the analytic computation presented in Fig. 9.1. That curve is shown in this figure, dotted and slightly displaced above the numerically computed curve. The two curves are identical.

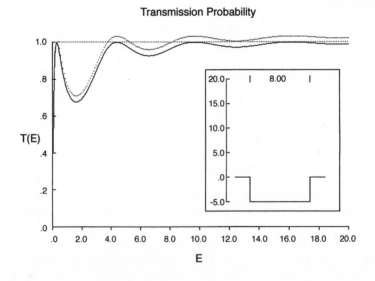

Fig. 10.2 Numerically computed transmission probability for the "attracting barrier" shown in the inset. The curve from the analytic computation (Fig. 9.2) is shown dotted and slightly offset. The two curves are identical.

11

Shape of Barrier

The plots of transmission probability versus energy given in Figs. 9.1–10.2 for a single rectangular barrier show a lot of structure. In particular, they exhibit peaks in the classical region ($E > V$) at which the transmission probability assumes the value +1. These peaks are suggestive of a "resonance structure" (see chapter 16).

We should wonder what part of the structure shown in the $T(E)$ versus E plots is intrinsic to quantum mechanical systems, and what part is an artifact of the particular potential shape chosen, with "square corners" and discontinuities between separate regions. To addess this question we study the transmission probability spectrum $T(E)$ for two smoother potentials.

The smooth scattering potential that we study has a Gaussian shape:

$$V(x) = V e^{-x^2}, \qquad -4 \leq x \leq +4 . \tag{11.1}$$

We study this potential by making a piecewise approximation to it. The interval from $x = -2$ to $x = +2$ is divided into N (odd) equal length intervals. The value of the piecewise constant potential in the jth interval ($1 \leq j \leq N$) is chosen to be the value of $V(x) = V e^{-x^2}$ at the midpoint of that interval.

In Fig. 11.1 we plot $T(E)$ for a five-piece approximation to this potential for $V = 5$ eV in the interval $0 < E \leq 20$ eV. The inset shows how well the piecewise constant potential approximates the smooth potential. We see from this plot that most of the structure above $V = 5$ eV has been washed out, but that some residual structure still remains.

In Fig. 11.2 we repeat the calculation shown in Fig. 11.1, but for an eleven-piece approximation to this smooth potential. Essentially all the structure above $E = V$ has been removed by this eleven-piece approximation to the smooth potential. The

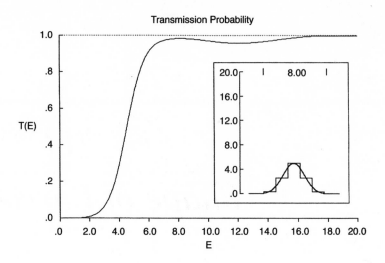

Fig. 11.1 Transmission probability for the five-piece approximation to the smooth repelling Gaussian potential shown in the inset.

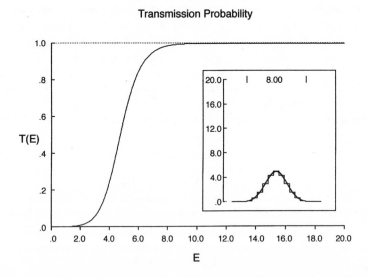

Fig. 11.2 Transmission probability for the eleven-piece approximation to the smooth repelling Gaussian potential shown in the inset. The resonance structure that appears in Figs. 9.1 and 10.1 for the one-piece approximation to the smooth potential is washed out in this smoother approximation.

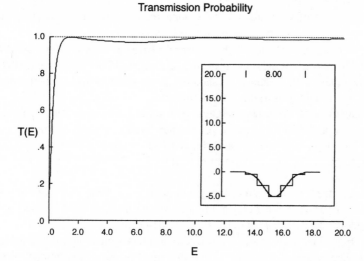

Fig. 11.3 Transmission probability for the five-piece approximation to the smooth attracting Gaussian potential shown in the inset.

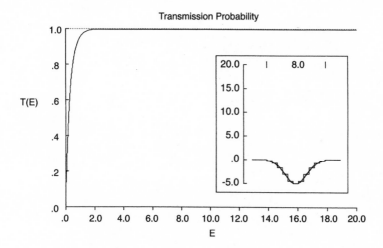

Fig. 11.4 Transmission probability for the eleven-piece approximation to the smooth attracting Gaussian potential shown in the inset. The resonance structure which appears in Figs. 9.2 and 10.2 for the one-piece approximation to the smooth potential is washed out in this smoother approximation.

$T(E)$ versus E curve does not show any additional changes as the approximation improves (to $N = 101$).

A similar set of computations was done for the attracting potential "barrier," $V(x) = Ve^{-x^2}, V < 0$. The results are shown for $V = -5$ eV, for five- and eleven-piece approximations to this inverted Gaussian potential, in Figs. 11.3 and 11.4. We again see that the resonance structure, apparent in Figs. 9.2 and 10.2 for the attracting square well, becomes washed out as we make better and better piecewise constant approximations to the smooth potential.

The calculations done here suggest that some of the features apparent in the transmission probability computed for a square well scattering or binding potential (Figs. 9.1 through 10.2) are artifacts due to the discontinuities between potentials in adjacent regions. As the discontinuities become smaller, the structures they produce also diminish.

However, there is one feature that is not an artifact of the shape of the potential. This is the depression of $T(E)$ below one for an attracting barrier that occurs at small energy. That is, the dip near $E = 0$ that occurs in Figs. 9.2 and 10.2 is still present (Figs. 11.3 and 11.4) in the five- and eleven-piece approximations to the inverted Gaussian potential. This is not an artifact of our computational procedure: this phenomenon is exhibited by real quantum mechanical systems.

12

Asymptotic Behavior

In this chapter we investigate the asymptotic behavior of the transmission probability under two different conditions. These conditions involve tunneling through a scattering potential and transmission through an attracting potential.

The arguments leading to equation (9.9) suggest that the transmission probability through a classically forbidden region behaves exponentially like $T(E) \sim e^{-2\kappa L}$, where L is the width of the potential. In Fig. 12.1 we plot $\ln [T(E)]$ versus L for the square well barrier shown in the inset. This clearly shows linear behavior in L for sufficiently large L ($L > 1$ Å). For this calculation, $V = 5.0$ eV, $E = 4.0$ eV.

In Fig. 12.2 we plot $\ln [T(E)/T(V)]$ as a function of $\kappa = \sqrt{2m(V-E)/\hbar^2}$ for $L = 1$ Å, $E = 4.0$ eV, where V is varied from $E + 0.0001$ eV to 105.0 eV. This also shows linear behavior in κ for κ sufficiently large ($\kappa > 2$ Å^{-1}).

These results suggest an asymptotic behavior for the tunneling probability, which has the form

$$
T(E) \quad \sim \quad \left| \exp - \int_a^b \kappa(x)dx \right|^2
$$

$$
= \quad \exp -2 \int_a^b \sqrt{2m(V(x) - E)/\hbar^2} \, dx \, . \tag{12.1}
$$

The integral is carried out through the classically forbidden region $V(x) - E \geq 0$, $a \leq x \leq b$.

In Fig. 12.3 we plot $- \ln T(E)$ versus $2 \int_a^b \kappa(x)dx$ for the Gaussian scattering potential shown in the inset. The calculations have been carried out for potentials of width $L = 1, 3, 5, 7, 9$ Å, of height $V = 5, 15, \cdots, 105$ eV, using $N = 1$-, 5-, 9-, 13-piece approximations to this potential. The incident particle energy was scanned in

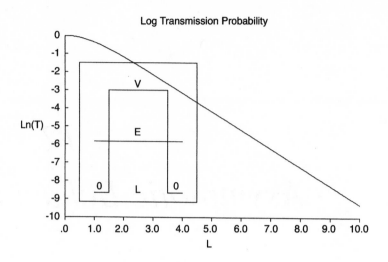

Fig. 12.1 Natural logarithm of the transmission probability decreases linearly with the width of the barrier, for sufficiently wide barriers. Parameter values for this plot: $V = 5.0$ eV, $E = 4.0$ eV.

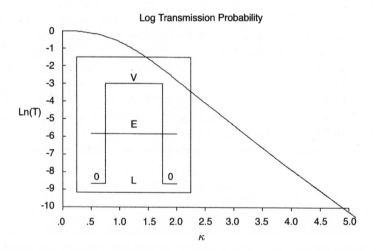

Fig. 12.2 Natural logarithm of the transmission probability decreases linearly with κ for sufficiently large values of κ. For this plot $E = 4.0$ eV, $L = 1.0$ Å, and the barrier potential is scanned from slightly above E ($V = E + 0.0001$ eV) to $V = 105.0$ eV. The transmission probability has been normalized by dividing by the transmission probability at the grazing energy, $T(E = V)$.

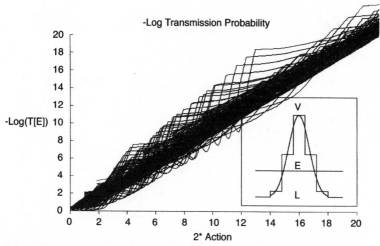

Fig. 12.3 Natural logarithm of the transmission probability is plotted against $\int_a^b \kappa(x)\ dx$ for a large number of approximations to the smooth potential shown.

the range of energies $+1$ eV $\leq E \leq (V-1)$ eV. The asymptotic energies on the left and right are $V_L = V_R = 0$. The sharp kinks are related to the passage of the scanning variable E through the discontinuities that are apparent in the piecewise constant approximation to the smooth potential. As N increases, the kinks become washed out.

The asymptotic properties of the transmission probability through an attracting barrier are simple to discuss. For most energies, $T(E) \sim 1$. However, for low energy ($E \sim 0$) the transmission probability exhibits structure that is not an artifact of our computational procedure.

To get some indication of this structure, we have computed $T(E)$ for the attracting Gaussian potential $V(x) = Ve^{-x^2}$ (-20 eV $< V < 0$ eV, -4 Å $\leq x \leq +4$ Å). The computation was done using a piecewise constant approximation to $V(x)$ involving $N = 101$ pieces. The calculations were done at many energies. We show the results for two energies, $E = 0.1$ eV (Fig. 12.4) and $E = 0.01$ eV (Fig. 12.5). The trends in behavior are apparent from these figures. As E increases, the structure exhibited in these two figures washes out (cf. also Figs. 11.3, 11.4). As the energy decreases, the peaks become narrower and the minima between them become deeper. As $E \to 0^+$ each peak becomes an infinitely thin spike and the transmission probability between successive spikes quickly approaches zero.

We will see later (Part III, chapter 25) that these peaks are intimately related to the formation of new bound states in potentials of finite depth.

Problem. Compute the width of the lowest peak ($V = -2.2$ eV) as a function of E.

Problem. Compute the asymptotic form of $T(E)$ as a function of E between the two lowest peaks (excluding the peak at $|V| = 0$) at about $V = -5.5$ eV.

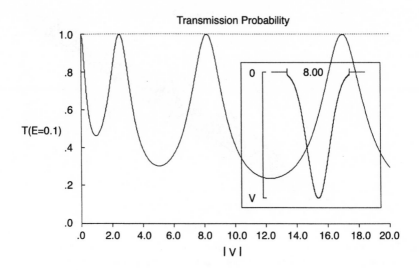

Fig. 12.4 Transmission probability for incident electron energy $E = 0.1$ eV for attracting potentials (inset) of various depth.

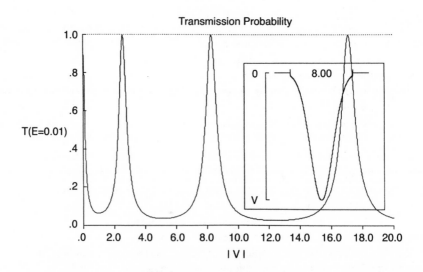

Fig. 12.5 Transmission probability for incident electron energy $E = 0.01$ eV for attracting potentials of various depth. The sharp peaks at low scattering energy are closely related to the bound states that this potential possesses.

13

Phase Shifts

Scattering phenomena exhibit another type of asymptotic behavior that is more subtle than tunneling, and which deserves to be discussed in a chapter of its own. This is the phenomenon of scattering phase shifts.

For a target for which $V_L = V_R = 0$, $T(E) = |1/t_{11}(E)|^2$. In general, $t_{11}(E)$ is a complex function that we can write as

$$t_{11}(E) = \frac{1}{\sqrt{T}}\, e^{-i\phi}\,. \tag{13.1}$$

For the single barrier of potential $V(< E)$ and width L, we have already computed $t_{11}(E)$ (see (9.5))

$$t_{11}(E) = \cos k'L - \frac{i}{2}\left(\frac{k'}{k} + \frac{k}{k'}\right)\sin k'L\,, \tag{13.2}$$

where $k = \sqrt{2mE/\hbar^2}$, $k' = \sqrt{2m(E-V)/\hbar^2}$. The angle $\phi(E)$ can be determined by taking the ratio of the imaginary to real values in (13.1) and (13.2):

$$\tan\phi = \frac{1}{2}\left(\frac{k'}{k} + \frac{k}{k'}\right)\tan k'L\,. \tag{13.3}$$

It is not entirely straightforward to solve this equation for ϕ. This comes about because $\tan(\phi + n\pi) = \tan\phi$ (n integer), so that many different values of ϕ can satisfy this equation. However, we recognize that as long as $k, k' \neq 0, \infty$

$$\begin{aligned}
\tan\phi = 0 &\iff \tan k'L = 0\,,\\
\cot\phi = 0 &\iff \cot k'L = 0\,.
\end{aligned} \tag{13.4}$$

Therefore, as we change parameter values (E, V, L) we can determine that ϕ is within $\pm \pi/2$ radians of $k'L$. In particular, we determine

$$\phi = \tan^{-1}\left(\frac{1}{2}\left(\frac{k'}{k} + \frac{k}{k'}\right) \tan k'L\right) + \pi\left[\frac{k'L}{\pi} + \frac{1}{2}\right] \qquad (13.5)$$

The complicated form of this expression comes about for two reasons:

1. The principal value of the $\tan^{-1}$ function is in the range $-\pi/2 < \phi < +\pi/2$.

2. The "greatest integer" function ($[x]$) truncates the decimal part of a real number x, rather than rounding to the nearest integer.

We show in Fig. 13.1 a plot of $\phi/2\pi$ against $k'L/2\pi$ for the potential shown in the inset. We expect $\phi(E)$ to intersect the diagonal ($\phi = k'L$) at integer values by (13.4). This in fact occurs. However, this plot holds two surprises:

1. $\phi(E)$ behaves asymptotically like $k'L$ even for relatively small values of $k'L$ ($k'L/2\pi > 2$).

2. $\phi(E)$ shows significant structure for $k'L$ small ($0 < k'L/2\pi < 1$).

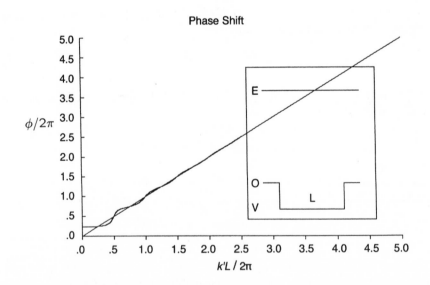

Fig. 13.1   Phase shift $\phi/2\pi$ of a particle with incident energy E as a function of $k'L/2\pi$, the action on traversing the barrier shown. For this calculation, $E = 100.0 \, \text{eV}$, $L = 4.0 \, \text{Å}$, and V is scanned from just below 100 eV to large negative values.

What does this phase mean physically? To answer this question, we compute ϕ when there is no scattering potential at all ($V = V_L = V_R = 0$). Then $k = k'$ and

$$
\begin{aligned}
t_{11}(E) &= \cos k'L - \frac{i}{2}\left(\frac{k'}{k} + \frac{k}{k'}\right)\sin k'L \\
&= \cos kL - i\sin kL = e^{-ikL} \; .
\end{aligned}
\tag{13.6}
$$

Thus, $\phi = kL$. This is exactly the change in phase that occurs as the particle moves from one side of the "potential" at $x = a_L$ to the other side at $x = a_R$ through a distance L, for

$$
\begin{aligned}
\Phi(x)|_{x=a_R} &= A\,e^{ika_R} = A\,e^{ik(a_R - a_L + a_L)} \\
&= e^{ikL}\,\Phi(x)|_{x=a_L} \; .
\end{aligned}
\tag{13.7}
$$

For more complicated potentials

$$
t_{11}(E) = \frac{1}{\sqrt{T}}\,e^{-i\phi} = \frac{1}{2}\left(m_{11} + \frac{k_R}{k_L}m_{22}\right) + \frac{i}{2}\left(k_R m_{12} - \frac{m_{21}}{k_L}\right) \; .
\tag{13.8}
$$

In a typical scattering experiment $V_L = V_R = 0$, $k_L = k_R = k_0$, and the phase shift is given, up to an integer multiple of π, by

$$
\phi = \tan^{-1}\left(\frac{-k_0 m_{12} + m_{21}/k_0}{m_{11} + m_{22}}\right) \; .
\tag{13.9}
$$

In the classically allowed regime where reflection can be neglected, the expression for the phase shift is well approximated by

$$
\phi = \int_a^b k(x)\,dx \; .
\tag{13.10}
$$

This can be approximated, for $E = \hbar^2 k_0^2/2m$, by

$$
\phi = \int_a^b \sqrt{2m(E - V(x))/\hbar^2}\,dx \simeq k_0 L\left(1 - \frac{\overline{V}}{2E}\right) \; .
\tag{13.11}
$$

Here the average energy is defined by $\overline{V} = \int_a^b V(x)dx/L$, with $L = b - a$. This shows that the phase shift over a length L is monotonically increasing and asymptotically approaches $k_0 L$ for sufficiently large energies.

In Figs. 13.2 and 13.3 we show the phase shift for an electron incident on an attracting and a repelling Gaussian potential $V(x) = V_0 e^{-(x/2)^2}$, $V_0 = \mp 5$ eV. An eleven-piece approximation to the potential is shown in the inset. The phase shift $\phi(E)$ is plotted both as a function of the parameter $k_0 L$ (curve a) and $\int_{-4}^{+4} k(x)dx$ for the attracting potential ($V_0 = -5$ eV), and the real part of this this integral for the potential barrier ($V_0 = +5$ eV) (curve b). Curves a and b approach each other asymptotically as E becomes sufficiently large.

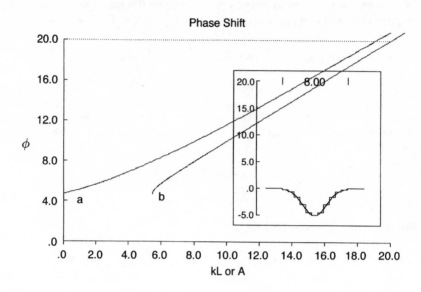

Fig. 13.2 Phase shift ϕ for the attracting barrier (inset) plotted as a function of (a) $k_0 L$ and (b) action integral $A = \int k(x)dx$. The phase shift at $E = 0$ is positive since the particle speeds up on passing through the potential.

We point out that the phase shift at zero energy is positive for the attracting potential and negative for the repelling potential.

Problem. Show that the tunneling results ($E < V$) and the phase shift results ($E > V$) can be put into the following similar form

$$t_{11}(E) \quad \sim \quad \exp{-i \int_a^b \sqrt{2m(E - V(x))/\hbar^2}\ dx}$$

$$= \quad \exp{-\frac{i}{\hbar} \int_a^b p(x)\ dx} , \qquad (13.12)$$

where $V(a) = V(b) = E, p(x) = \sqrt{2m(E - V(x))}$ and the appropriate square root is taken when $E - V(x) < 0$. The integral $A = \int_a^b p(x)\ dx$ is called the *classical action* of the particle.

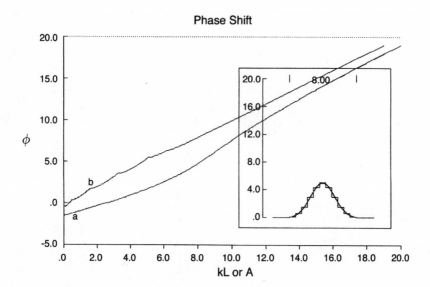

Fig. 13.3 Phase shift ϕ for the repelling barrier (inset) plotted as a function of (a) $k_0 L$ and (b) action integral $A = \int k(x)dx$. The phase shift at $E = 0$ is negative since the particle slows down on passing through the potential.

14

Double Barrier

Up to now we have considered tunneling problems for which the classically forbidden region is one contiguous region. In such cases we have seen that the transmission probability has an asymptotic form

$$T(E) \sim \exp\left[-2 \int_a^b \sqrt{\frac{2m(V(x) - E)}{\hbar^2}} \, dx\right] . \qquad (14.1)$$

This asymptotic form is no longer accurate when the classically forbidden region consists of two or more disjoint pieces. Then resonances can occur between adjacent classically forbidden regions that alter $T(E)$ in a very significant way.

We illustrate what can occur by computing the transmission probability for the double well potential shown in Fig. 14.1 (inset). The remarkable feature to be observed is that the transmission probability is very large, in fact +1, for certain values of the energy for which transmission is classically forbidden. The potential shown has two barriers of height $V = 5$ eV and width $D = 2$ Å separated by a region at 0 eV and width $L = 6$ Å; $V_L = V_R = 0$. It appears that the lowest peak does not reach $T(E) = 1.0$. However, this is a resolution problem. The peak is so narrow that the energies for which $T(E)$ was computed (every 0.002 eV) only sampled the shoulder of this peak.

A number of questions should naturally be asked. These include:

- Why does this phenomenon occur?

- How are the location and width of the peaks related to the parameters of the potential?

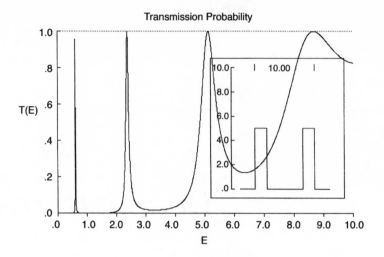

Fig. 14.1 Transmission probability spectrum for double barrier scattering potential shown in inset. The two identical barriers have width 2 Å and height 5 eV. They are separated by 6 Å. The peak at 0.6 eV is not completely resolved at the energy resolution (0.002 eV) used for this computation.

- What is the shape of the peaks?

We will answer these questions in reverse order. Briefly, we show in the remainder of this chapter that the peaks have a Lorentzian line shape. We leave it to the first problem at the end of this chapter to show that the peaks occur at energies $E_n \sim \frac{\hbar^2}{2m} \left(\frac{n\pi}{L}\right)^2$, where L is the width of the intermediate classically allowed region. In the second problem we show that the width of the peaks decreases exponentially with the thickness, D, of the classically forbidden region. In chapter 16 we show that this phenomenon is due to the occurrence of resonances within the classically allowed regions.

In order to show that the peak at $E \sim 0.6$ eV in Fig. 14.1 really rises to $T(E) = 1$ for some value of E, we have resolved the peak by scanning the energy from $E = 0.59$ eV to $E = 0.64$ eV in 500 steps. The shape of this lowest resonance is shown in Fig. 14.2. This clearly has a maximum at $T(E) = 1$ for $E \simeq 0.615$ eV.

Two reasonable candidates for describing the bell-shaped curve are the Gaussian function and the Lorentzian function. We could try to fit the data in Fig. 14.2 to each type of curve and then determine how good or bad the fit is. This is a standard problem of statistics. We will not pursue this approach here. Rather, we will determine the consequences of each functional form and compare these consequences to the data.

Gaussian. If the curve shown in Fig. 14.2 is a Gaussian, it has the form

$$T(E) = Ae^{-[(E-E_0)/\Delta E]^2} \, , \tag{14.2}$$

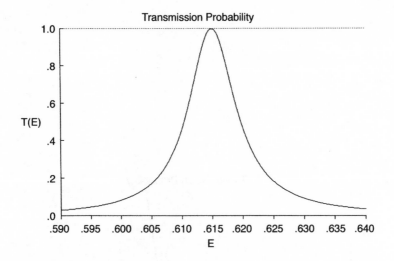

Fig. 14.2 Lowest peak in the transmission spectrum shown in Fig. 14.1 is resolved by a higher resolution ($\Delta E = 0.0001$ eV) energy scan.

where E_0 is the location of the peak and ΔE is related to its half width. By taking the negative logarithm we should have a rising parabola

$$- \ln T(E) = \left(\frac{E - E_0}{\Delta E}\right)^2 - \ln A . \tag{14.3}$$

The first derivative is a linear function, and the second is a constant:

$$
\begin{aligned}
\frac{d}{dE}(-\ln T) &= 2\left(\frac{E - E_0}{\Delta E}\right)\frac{1}{\Delta E} , \\
\frac{d^2}{dE^2}(-\ln T) &= \frac{2}{\Delta E^2} .
\end{aligned}
\tag{14.4}
$$

Fig. 14.3 repeats the plot of $T(E)$ versus E shown in Fig. 14.2, shows $-\ln T(E)$ versus E (looks like a parabola) and shows a plot of the first derivative of this function. The first derivative is definitely not a straight line. Therefore we can reject the guess that the resonance has a Gaussian form.

Lorentzian. If the curve shown in Fig. 14.2 is a Lorentzian, it has the form

$$T(E) = \frac{A}{1 + \left(\frac{E - E_0}{\Delta E}\right)^2} , \tag{14.5}$$

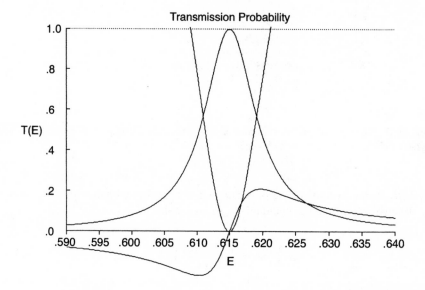

Fig. 14.3 Negative logarithm of the transmission peak (parabola-shaped curve) superposed on the peak. If the peak is a Gaussian, the derivative of its negative logarithm will be a straight line. The derivative is not even close to a straight line, so the peak is not approximated by a Gaussian function.

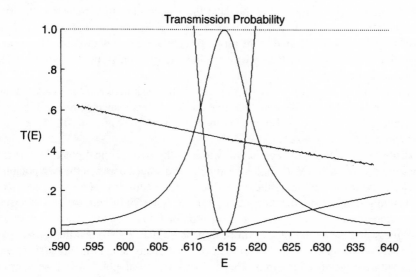

Fig. 14.4 Reciprocal of the transmission peak (parabola-shaped curve) superposed on the peak. If the peak is a Lorentzian, the derivative of the reciprocal will be a straight line with positive slope and the second derivative will be a constant. The first derivative is well approximated by a straight line with positive slope. The second derivative (with negative slope) does not vary "too much" through this resonance. This shows the resonance curve is well approximated by a Lorentzian function with energy maximum at 0.6150 eV and half width (at half height) 0.0046 eV.

where E_0 is the location of the peak and ΔE is the half width at half height. By taking the reciprocal we should find a rising parabola

$$\frac{1}{T(E)} = \frac{1}{A} + \frac{1}{A}\left(\frac{E - E_0}{\Delta E}\right)^2. \tag{14.6}$$

Once again, the first derivative should be a linear function of E with positive slope, and the second derivative should be a constant. In Fig. 14.4 we show the resonance, with its reciprocal -1 (since $A = 1$), the first, and the second derivative. These plots are superposed on the original resonance curve. The first derivative (positive slope) is computed by first differencing the data; the second derivative (negative slope) is the first difference of the first derivative (i.e., second difference). The scales of these derivatives have been adjusted so the properties of the curves are evident. The second derivative decreases somewhat with increasing energy. This is an effect of the peaks at higher energies.

The results of Figs. 14.3 and 14.4 clearly distinguish between the Gaussian and Lorentzian line shapes. We conclude that the lowest resonance shown in Fig. 14.1 and enlarged in Fig. 14.2 is Lorentzian in shape, with peak at $E = 0.6150$ eV and half width $\Delta E = 0.0046$ eV.

Problem. Fix V, D, the height and width of the two identical potential barriers (for example, $V = 20$ eV, $D = 2$ Å). Vary L, the distance between the two potential barriers, and try to determine the dependence of the centers of the lowest peaks on L. You might infer from the data that $E_c(n)$, the center of the nth peak, is inversely proportional to L^2. Plot $E_c(n)L^2$ as a function of L to check this guess. You might also guess that the center of the nth peak behaves like n^2. To test this guess, plot $E_c(n)L^2/n^2$ versus L for the lowest peaks in the transmission probability spectrum. The results are shown in Fig. 14.5. This figure shows: the double barrier; the energies at which the transmission peaks occur for $E < V = 20$ eV, plotted as a function of L; and the ratio of the transmission peak energies, $E/E(n, L)$, plotted as a function of L, where $E(n, L) = (\hbar^2/2m)(n\pi/L)^2$.

Problem. Fix L and V, and vary D (for example, $V = 20$ eV and $L = 4$ Å). You might expect the half widths of the Lorentzian peaks to decrease exponentially with the width D of the classically forbidden region: $\Delta E \sim e^{-\lambda D}$. To test this hypothesis, plot $-\ln(\Delta E)/D$ as a function of D for the lowest transmission resonance peaks. In fact, you might even guess that $-\ln(\Delta E) \sim \int \kappa(x)\,dx$, where the integral extends through (both) classically forbidden regions. To test this guess, plot $-\ln(\Delta E)/(2D\sqrt{2m(V - E_{\text{peak}})}/\hbar^2)$ as a function of D for the lowest peaks in the transmission spectrum. The results are shown in Fig. 14.6. This figure shows: the double barrier ($V = 20$ eV and $L = 4$ Å); the natural logarithm of the full width at half height of the Lorentzian peak plotted as a function of D, the width of either barrier; and the ratio of this logarithm to the action, $-\ln(\Delta E)/A$, plotted as a function of D, where $A = 2D\sqrt{2m(V - E_{\text{peak}})}/\hbar^2$.

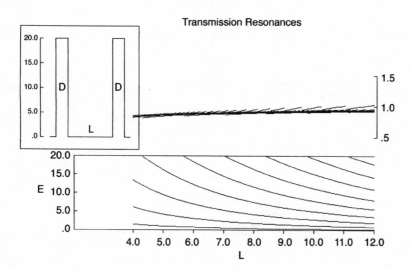

Fig. 14.5 Plot of $E_n/E(n, L)$ vs. L for the double barrier potential, where $E(n, L) = (\hbar^2/2m)(n\pi/L)^2$. The two identical barriers have constant height 20 eV and thickness 2 Å, but their separation, L, varies from 4 Å to 12 Å. The number of resonances below 20 eV varies from three at L = 4 Å to nine at 12 Å. The approach of all scaled energies to a common value suggests that for the "deep" resonances ($\kappa L/\hbar \geq 1$) the energies of the resonances are $E_n \simeq (\hbar^2/2m)(n\pi/L)^2$.

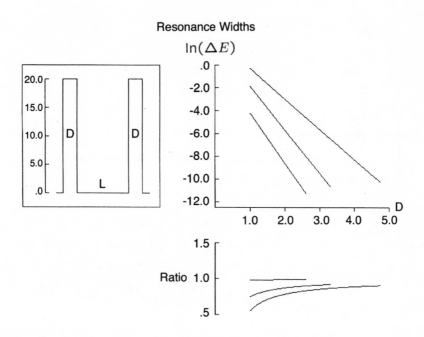

Fig. 14.6 Plot of $-\ln(\Delta E)$ as a function of D for the double barrier potential, where ΔE are the full widths at half height of each of the three resonance peaks that exist for this double barrier potential. The two identical barriers have constant height 20 eV and separation 4 Å, but their thickness, D, varies from 1.0 Å to 5.0 Å. *Below*, Plot of the ratio $-\ln(\Delta E)/2\kappa D$ as a function of D, where $\kappa = \sqrt{2m(V - E_{\text{peak}})/\hbar^2}$. The approach of all scaled ratios to a common value suggests that the linewidth of the "deep" resonances decreases exponentially: $\Delta E \simeq \exp\left[-2D\sqrt{2m(V - E_{\text{peak}})/\hbar^2}\right]$.

15

Multiple Barriers

If a second barrier behind the first produces surprises in the transmission probability spectrum (see Fig. 14.1), what will a third behind the second do?

To explore this question we computed $T(E)$ for a potential consisting of three rectangular barriers, each of height V eV and width D Å, and separated from each other by a distance L Å (Fig. 15.1). The principal effect, clearly visible in the plot of $T(E)$ versus E, is that each peak in the transmission spectrum of the double barrier splits into a doublet in the transmission spectrum for the triple barrier.

The transmission probability spectrum for a quadruple barrier, formed as described above but with one more rectangular barrier, is shown in Fig. 15.2. Each peak is now split into a triplet of peaks.

This behavior is systematic. For a potential consisting of $N + 1$ identical barriers separated by equal distances, we find

1. Each peak splits into an N-tuplet.

2. The "center of gravity" of each N-tuplet occurs at roughly the same energy as the corresponding peak in the transmission spectrum of the double barrier.

3. The width of each multiplet grows slowly with N, and saturates for relatively small values of N. That is, the N-tuplet corresponding to one peak of the double barrier spectrum does not overlap the N-tuplet corresponding to another peak, at least for classically forbidden energies.

It is possible to determine how the width of each N-tuplet saturates as $N \rightarrow \infty$, but we delay answering this question until we discuss energy bands for periodic lattices (Part IV).

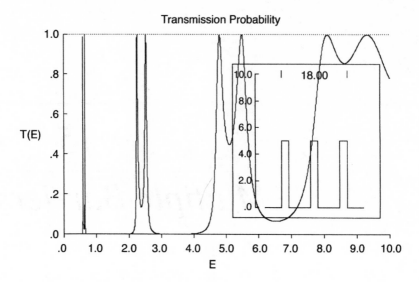

Fig. 15.1 Transmission probability spectrum for three identical barriers with $V = 5$ eV, $D = 2$ Å, and $L = 6$ Å. Each peak is a doublet.

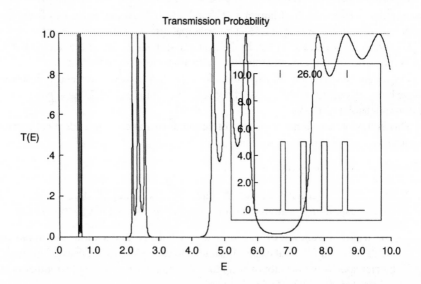

Fig. 15.2 Transmission probability spectrum for four identical barriers separating three identical wells. As in Fig. 15.1, $V = 5$ eV, $D = 2$ Å, and $L = 6$ Å. Each peak is a triplet.

16

Probability Distributions

In order to determine why there are resonances in the transmission probability spectrum of the double barrier, it is useful to compute the probability density of the electron as it is scattered from this potential.

The problem has been solved in principle in Part I, chapter 3. In that chapter we computed the transfer matrix relating amplitudes $\begin{bmatrix} A \\ B \end{bmatrix}_j$ in region j to the amplitudes $\begin{bmatrix} A \\ B \end{bmatrix}_{j+1}$ in region $j + 1$. These 2×2 transfer matrices (3.7) are complex. It is possible to reduce the number of complex operations involved in computing the probability density.

To do this, we choose a different set of solutions to Schrödinger's equation in regions of constant potential:

$$
\begin{array}{cccc}
 & \Phi_1(x) & \Phi_2(x) & \\
E > V & \cos kx & \sin kx & k = \sqrt{2m(E - V)/\hbar^2} \\
E = V & 1 & x & \\
E < V & \cosh \kappa x & \sinh \kappa x & \kappa = \sqrt{2m(V - E)/\hbar^2} \, .
\end{array}
\tag{16.1}
$$

In addition, within each region we measure distance from the left edge to the right $(0 \le x_j \le \delta_j$ for x_j in region $j)$. Imposing continuity of the wavefunction and its first derivative at the boundary between region j and region $j + 1$ leads to an equation of the form

$$
\begin{bmatrix} \cos k_j \delta_j & \sin k_j \delta_j \\ -k_j \sin k_j \delta_j & k_j \cos k_j \delta_j \end{bmatrix} \begin{bmatrix} A \\ B \end{bmatrix}_j = \begin{bmatrix} 1 & 0 \\ 0 & k_{j+1} \end{bmatrix} \begin{bmatrix} A \\ B \end{bmatrix}_{j+1}
\tag{16.2}
$$

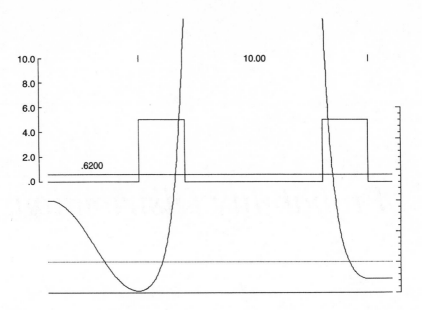

Fig. 16.1   Probability density for an electron of energy $E = 0.62$ eV incident on the double barrier shown in Fig. 14.1. The energy chosen sits on the shoulder just above the peak, at about $T(E) = 0.6$ (see Fig. 14.2). Closer to the peak, the probability density between the barriers becomes much larger.

in the case $E > V_j, E > V_{j+1}$. Analogous equations apply for all conditions (i.e., $E > V_j, E < V_{j+1}$, etc.). All 2×2 transfer matrices involved in propagating the amplitudes $\begin{bmatrix} A \\ B \end{bmatrix}_{j+1}$ to the amplitudes $\begin{bmatrix} A \\ B \end{bmatrix}_j$ in the adjacent region are now real. Only the amplitudes themselves are complex.

This set of equations for the amplitudes is initialized by setting

$$\Phi_R(x) = \sqrt{T}\, e^{ik_R x} = \sqrt{T}\, (\cos k_R x + i \sin k_R x)\,,$$

$$\begin{bmatrix} A \\ B \end{bmatrix}_R = \sqrt{T} \begin{bmatrix} 1 \\ i \end{bmatrix}. \tag{16.3}$$

Using this procedure, we have computed the probability density $|\Phi(x; E)|^2$ for a particle of energy E scattered by the double well potential shown in Fig. 14.1. These probability densities are computed for energies near the maxima ($E = 0.62$ eV, $E = 2.31$ eV) of the lowest two peaks, and are shown in Figs. 16.1 and 16.2.

The results of these computations can be summarized in the following observations, some of which are apparent from the figures.

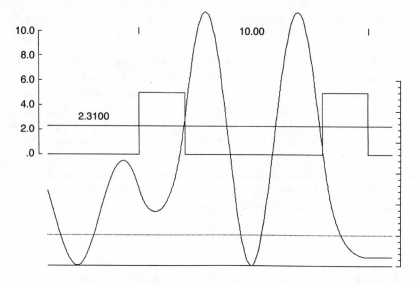

Fig. 16.2 Probability density for an electron in the double barrier potential shown in Fig. 14.1. The energy ($E = 2.31$ eV) is just below the second peak. At this energy, the transmission probability is about 0.28.

1. The probability density in the right-hand region is constant. This follows from the boundary condition that is imposed: $\Phi_R(x) = \sqrt{T}\, e^{+ik_R x}$, $|\Phi_R(x)|^2 = T$.

2. The probability density oscillates in the left-hand region. This also follows from the boundary condition: $\Phi_L(x) = e^{+ik_L x} + Re^{-ik_L x}$, $|\Phi_L(x)|^2 = 1 + \overline{R}R + (Re^{-2ik_L x} + \overline{R}e^{+2ik_L x})$. The wavelength of the oscillation is determined by k_L ($\lambda = \pi/k_L$); the phase of R determines the probability at the left-hand edge of the scattering potential. If $R = 0$ (i.e., $T = 1$), the probability density in the left-hand region is also constant and equal to the constant value in the right-hand region if $V_L = V_R$.

3. The probability density between the double barriers can be very high. The ratio of the maximum probability density between the barriers to the incident intensity varies as the energy is swept and assumes a maximum value at the resonance peaks.

4. Near the lowest peak of the transmission spectrum $T(E)$ the probability density has a single maximum. Near the second peak in the transmission spectrum $T(E)$, the probability density exhibits two peaks. These peaks are separated by a node that approaches zero quadratically.

5. The probability density associated with the nth peak in the spectrum of $T(E)$ shows n peaks. These are separated by $n-1$ nodes that approach zero quadratically.

6. Peaks in the transmission spectrum occur at energies for which particles exhibit constructive interference. That is, when a particle inside the double barrier starts at the right-hand edge of the barrier on the left and travels to the left-hand edge of the barrier on the right, it undergoes a phase shift $\int_a^b p\ dx/\hbar = kL$. On reflecting off the right-hand barrier, it undergoes a phase shift of approximately π (the larger $\sqrt{2m(V-E)/\hbar^2}\ D$, the closer to π). On traveling to the left-hand well and reflecting off it, the particle undergoes another equal phase shift of about $kL+\pi$. If the total phase shift in this round trip is an integer multiple of 2π, the particle will interfere with itself constructively. Therefore, resonances occur for $2kL + 2\pi \sim 2\pi n$, or energies given by

$$E_n \sim \frac{\hbar^2}{2m}\left(\frac{n\pi}{L}\right)^2 . \tag{16.4}$$

This approximation is valid when $\sqrt{2m(V-E)/\hbar^2}\ D$ is large.

The probability densities for triple, quadruple, and so forth, barrier potentials have also been computed. For a triple barrier, each peak in the transmission spectrum is a doublet (see Fig. 15.1). The probability densities for energies corresponding to the two peaks in a doublet are very similar. We will not learn what the difference between the corresponding scattering states is by studying probability densities. Rather, we must study the appropriate wavefunctions. This study will be left to Part III, chapter 29.

17

Combining Barriers

In the previous chapters, we have developed some understanding of quantum mechanical tunneling through a single barrier. We have also studied tunneling through double (and multiple identical) barriers and have understood the occurrence of peaks in the transmission probability spectrum as a resonance phenomenon.

We now ask what happens when we combine several different barriers. To this end, we study the triple barrier system consisting of barriers A ($V = 5.0$ eV, $\delta = 1.5$ Å), B (4.0 eV, 2.0 Å), and C (3.0 eV, 2.5 Å). Barriers A and B are separated by 7 Å at $V = 0$ eV, while barriers B and C are separated by 3 Å, also at $V = 0$ eV.

Before studying this set of three barriers, we study separately the transmission probability spectrum for the double barriers AB (Fig. 17.1) and BC (Fig. 17.2). These plots contain no surprises. Peaks occur in the classically forbidden regime. For the double barrier AB (Fig. 17.1) these peaks occur at energies of about 0.4, 1.9, and 4.0 eV. These three peaks have maxima at $T(E) = 1.0$; the very narrow peak at 0.4 eV does not appear to have maximum at $T(E) = 1.0$ only because of resolution limitations. A peak occurs in the scattering region [$E > \min(A, B) = 4.0$], but this peak has a maximum transmission probability $T(E) < 1$.

The spectrum of $T(E)$ for the double barrier BC exhibits similar properties. One maximum occurs in the classically forbidden region [$E < \min(B, C) = 3.0$ eV] at about 1.3 eV. A second maximum occurs in the classically allowed region at about 4.5 eV, but at this maximum $T(E) < 1$.

The transmission probability for the triple barrier ABC is shown in Fig. 17.3. This spectrum exhibits peaks at about 0.4, 1.3, 1.9, 3.9, 4.6, and 7.0 eV. The peak at 0.4 eV is invisible due to resolution limits. In this figure we observe two things:

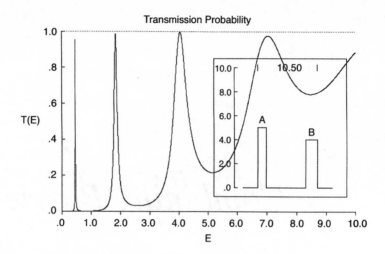

Fig. 17.1 Transmission spectrum of the double barrier AB. Transmission peaks occur in the classically forbidden regime at ~ 0.4, 1.9, and 3.9 eV. These have maxima at $T(E) = +1$, although they are resolution limited in this plot. The peak at 7.0 eV in the classically allowed regime has maximum at $T(E) < +1$. Inset provides information on the height and total length of the barrier.

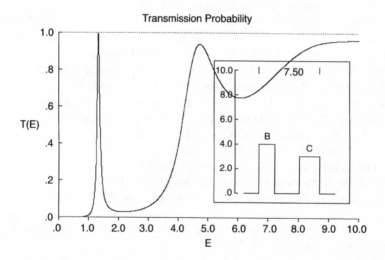

Fig. 17.2 Transmission spectrum of the double barrier BC. Transmission peaks at 4.5 and 10^+ eV in the classically allowed regime have maxima below $T(E) = +1$.

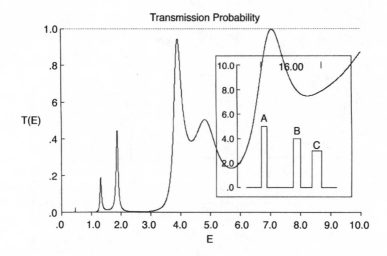

Fig. 17.3 Transmission spectrum of the double barrier ABC, A = (5.0, 1.5), B = (4.0, 2.0), and C = (3.0, 2.5) (eV, Å). Barriers A and B are separated by 7.0 Å, B and C by 3 Å. Peaks at 0.4, 1.9, 3.9, and 7.0 eV arise from resonances in the AB double barrier, those near 1.3 and 4.9 eV are due to resonances in the BC double barrier.

1. The peaks at 0.4, 1.9, 3.9, and 7.0 eV seem to be related to the double barrier AB (see Fig. 17.1), while the peaks at 1.3 and 4.6 eV seem to be related to the double barrier BC (see Fig. 17.2).

2. None of the peaks has a maximum value of +1.

We can check that the fingerprints we have used to identify the peaks that appear in Fig. 17.3 actually lead to correct identifications by computing relative probability densities for energies near these peaks. Probability densities in the triple barrier potential have been computed for $E = 1.31$ eV (Fig. 17.4), $E = 3.65$ eV (Fig. 17.5), and $E = 7.0$ eV (Fig. 17.6).

Fig. 17.4 ($E = 1.31$ eV) shows that the electron probability density is much larger between wells B and C than between A and B. This peak is therefore due to a resonance between wells B and C. Since an electron with energy matched to resonate in the BC well will not be in resonance in the AB well, the maximum of the transmission resonance peak near 1.3 eV is less than +1.

The probability distribution in Fig. 17.4 tells us more. Since the probability density in the BC well has no nodes, this peak corresponds to the lowest energy transmission peak in the BC double well potential. In particular, this means that there are no peaks below 0.4 eV in the transmission spectrum of the BC double barrier (Fig. 17.2), which are so narrow that they are not even hinted at in the transmission spectrum.

Fig. 17.5 ($E = 3.65$ eV) shows that the electron probability density in the well between barriers A and B is much larger than between barriers B and C. Thus the

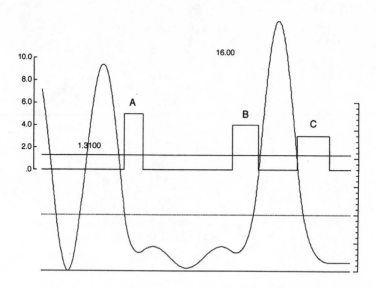

Fig. 17.4 *Above*, Triple well potential with energy scale at left. *Below*, Probability density for electron with energy $E = 1.31$ eV. *Dotted*, Probability density $= +1$ (scale at right). The resonance peak near 1.3 eV is due to the lowest resonance in the BC double barrier.

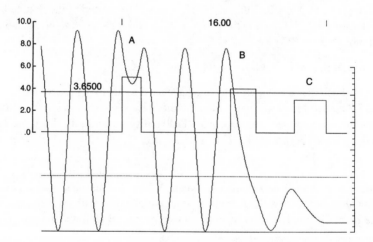

Fig. 17.5 *Above*, Triple well potential with energy scale at left. *Below*, Probability density for electron with energy $E = 3.65$ eV. *Dotted*, Probability density $= +1$ (scale at right). The resonance peak near 3.9 eV is due to the third resonance in the AB double barrier. The wavefunction has two nodes in the AB double barrier, but since the minimum probability density in the BC double barrier is not zero, the wavefunction has no nodes in this region.

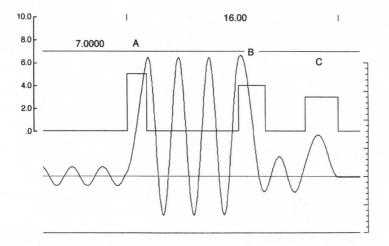

Fig. 17.6 *Above*, Triple well potential with energy scale at left. *Below*, Probability density for electron with energy $E = 7.0$ eV. *Dotted*, Probability density $= +1$ (scale at right). The resonance peak near 7.0 eV is due to the fourth resonance in the AB double barrier. The probability density for this classically allowed energy has minima that are nonzero, so the wavefunction has no nodes in this potential.

peak at 3.9 eV is primarily due to a resonance in the region between A and B. Which resonance? Since the probability density exhibits two nodes in this region of space, the peak is the third resonance (transmission peak) in the transmission spectrum of the AB double barrier.

The presence of one node between barriers B and C means that, at this energy between one and three $[(1+1) \pm 1]$ resonance peaks in the spectrum have occured due to resonances in the BC cavity. The location of the node can be used to determine whether this number is one, two, or three. We will not discuss this point now.

Fig. 17.6 ($E = 7.0$ eV) shows the electron probability density is larger between barriers A and B than between B and C. Thus, this peak is due to a resonance in the cavity between barriers A and B. Since the probability density has three minima in this region, it corresponds to the fourth transmission resonance of the AB cavity. We observe that the probability distribution has minima that do not reach zero. This means the wavefunction is not zero (there are no nodes) anywhere in this multiple barrier. This is characteristic of wavefunctions for electrons with energies in the classically allowed regime.

18

Quantum Engineering

As the size of electronic devices shrinks, the laws of quantum mechanics play an increasingly important role in their behavior. It should be possible to use these laws to design devices to operate within preset design specifications.

We illustrate this process with a simple example. Suppose electrons are conveniently available in some particular range of energies (e.g., 0-2 eV), but we need electrons in a much smaller energy range (e.g., 1 ± 0.2 eV). Can we design a filter that will pass electrons in this restricted range and reject electrons outside this range? How?

The specifications just described can be expressed as conditions on the transmission probability function $T(E)$: it is zero in the interval from 0 to 2 eV, except in the smaller interval around 1 eV from $1 - 0.2$ to $1 + 0.2$ eV. Our experience with transmission probability coefficients is

1. A single rectangular barrier will not exhibit resonance structures in the classically forbidden region.

2. A double barrier will exhibit resonance peaks at which 100% transmission is achieved.

3. Multiple different barriers will show resonance structure, but transmission peaks do not rise to $+1$.

4. Multiple identical equally spaced barriers produce multiplets of transmission peaks showing 100% transmission.

These observations suggest that a filter with the specified characteristics can be produced by fabricating a device with a large number of identical equally spaced barriers. If the barriers are rectangular, the multiple barrier device is specified by four parameters: $N + 1$, the number of barriers; V (eV), the height of each; D (Å), the width of each; and L (Å) the spacing between adjacent barriers.

A useful approach is to design a double barrier so that one of the resonance peaks falls more or less in the middle of the range of energies to be transmitted, while all other resonances fall outside this range. For the design characteristics specified, we will search for parameters L, D, V for which the lowest transmission peak is ~ 1 eV and the higher energy peaks occur for energies greater than 2 eV. The location of the transmission peaks depends more sensitively on L than D or V. We can use the (Action) resonance condition for round trips in the double barrier potential

$$2pL/\hbar + \text{phase shifts at 2 boundaries} = 2\pi n$$

to estimate L. The argument is that each reflection phase shift is $\sim \pi$, so that $2L\sqrt{2mE}/\hbar \sim 2\pi n$ or $L \sim n\pi/\sqrt{2mE/\hbar^2} \sim 6$ Å for $E = 1$ eV, $n = 1$. This quick and dirty estimate at least puts us in the right ballpark for a more refined estimation of the design parameters. After a few computations, we find that design parameters $L = 4.0$ Å, $D = 1.0$ Å, $V = 3.8$ eV produce a transmission probability spectrum, shown in Fig. 18.1, with the desired characteristics. That is, the lowest resonance occurs at about 1 eV and the next resonance is well above 2 eV.

In Fig. 18.2 we present the spectrum of $T(E)$ for a six-barrier device with five potential wells between the six barriers. Each peak from the double barrier potential is now a 5-plet. In each multiplet the maximum value of $T(E)$ is $+1$ and the minimum rises towards the center of the transmission band.

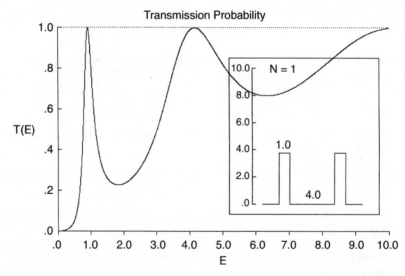

Fig. 18.1 Double barrier designed to have one peak at about 1.0 eV. *Inset*, Barrier height $V = 3.8$ eV, width $D = 1.0$ Å, separation between barriers $L = 4.0$ Å. There is one ($N = 1$) well between the two barriers.

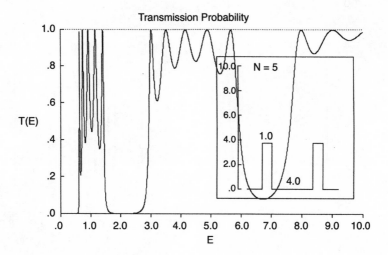

Fig. 18.2 Transmission probability spectrum of multiple barrier with N ($= 5$) wells formed by $N + 1$ barriers. N is given in the inset.

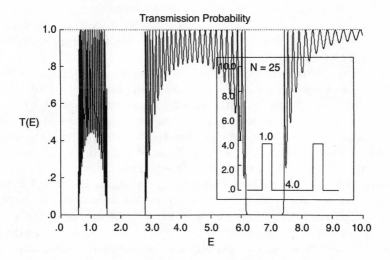

Fig. 18.3 Transmission probability spectrum of multiple barrier with $N = 25$. As the number of barriers (wells) increases, the pass bands approach $T(E) \simeq 1$ and the transmission probability between these bands approaches zero ("forbidden bands").

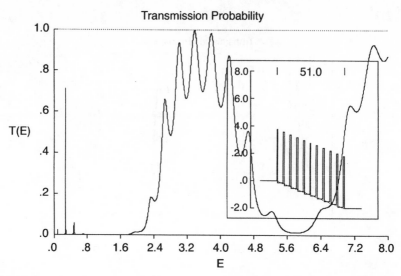

Fig. 18.4 Transmission probability spectrum for multiple barrier ($N = 10$) biased by a 2-V external electric field. The field displaces and distorts the transmission probability spectrum.

One way to smooth the transmission spectrum within each band is to crowd more peaks into this range. This is easily done by building more barriers. It is surprising, but true, that the width of each band is almost independent of N for N sufficiently large ($N \geq 5$). The width of these bands is strongly determined by D and V, more specifically by $D\sqrt{V - E_{Res}}$. We present the transmission spectrum of a device with $N = 25$ (26 barriers) in Fig. 18.3. Notice that the widths of the transmission pass bands are almost unchanged from the $N = 5$ to the $N = 25$ device.

The transmission properties of a multiple barrier potential can be altered by biasing it. That is, we impose an electric field on the device by creating a potential difference across it. In Fig. 18.4 we show (inset) an eleven-barrier potential with a 2-V bias. That is, the left-hand edge is grounded ("grounded" means the potential is zero), while the right-hand edge is held at +2 V. Since the electron charge is negative, the potential of the electron at the right-hand edge is -2 eV. We assume a linear decrease of potential between the left- and right-hand edges and approximate the barrier potential in the region as before. That is, we assume the potential in each of the $2N + 1$ regions is a constant whose value is the value of the potential at the midpoint of the region. Biasing in this way modifies the structure of the pass bands and pulls them to lower energy. If we assume that the electrons that are available to transit this barrier from left to right are thermal (i.e., energies $\leq 1/40$ eV at room temperature), then we can compute the transmission probability, and the conductance, of this device as a function of the bias voltage. The conductance function will not be a monotonic function of V and will therefore show regions of negative resistivity. A more complete discussion of this phenomenon is outside the scope of our subject, which is *elementary* quantum mechanics in one dimension.

19

Variations on a Theme

The theme of this part of the book has been scattering, with an emphasis on tunneling through one or more barriers. We have seen that peaks in the transmission probability spectrum occur even in the classically forbidden regime. These peaks are due to resonances.

We now ask: What type of phenomena can be anticipated if we are able to fabricate two different types of barriers and place them adjacent to each other in any desired order? To study this problem we introduce two simple rectangular barriers A and B, shown in the inset of Fig. 19.1. We choose A to be a repelling barrier with energy 5.0 eV and width 1.0 Å surrounded on each side by regions of zero potential and width 2.0 Å each. Barrier B has a similar shape, with height 3.5 eV and width 2.0 Å, surrounded by two regions of width 3.0 Å at 0 eV.

In order to interpret the spectrum of the barrier AABB shown in Fig. 19.1, it is useful first to determine the principal features of the three building blocks AA, AB, and BB. The principal features of these spectra are the locations of the peaks. The energies of the peaks below 10 eV are collected in Table 19.1.

The peak at 4.7 eV consists of two overlapping resonances, one from the double barrier AA and one from the double barrier BB. This peak cannot be resolved by increasing the resolution of the scan.

To be more precise, the somewhat distorted peak at 4.7 eV, which consists of two overlapping peaks, cannot be resolved by probing with increased energy resolution, *subject to the boundary conditions specified:* $V_L = V_R = 0$. In computing transmission spectra for double and multiple potential barriers, we have observed that the lower energy peaks are narrower than the higher energy peaks. This suggests that we might be able to resolve the overlapping resonances buried in the peak at 4.7 eV

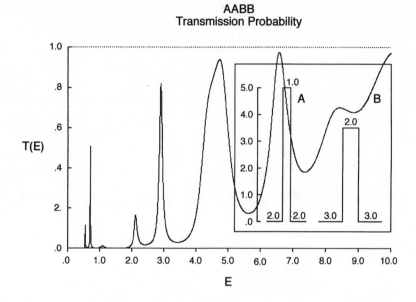

Fig. 19.1 Transmission probability for the multiple barrier AABB. The potentials A and B are shown in the inset. Each feature can be identified with a peak in one of the double barrier potentials AA, AB, or BB. The barriers AA and BB each contribute a peak in the range 4–5 eV. These peaks cannot be resolved by refining the energy scan.

by narrowing them. This can be done if we could somehow lower their energy. One possible way to do this would be to place this barrier inside a potential well with a depth of ~ 4 eV. To be more explicit, we could impose a potential on the asymptotic regions by setting $V_L = V_R = V$ and probe the target with an external low-energy electron beam with energy ϵ above the asymptotic limits V. This would probe the original potential (with $V_L = V_R = 0$) at the energy $E = V + \epsilon$.

The result of this kind of variable potential, low-energy spectroscopy is shown in Fig. 19.2 for $\epsilon = 0.1$ eV. The two peaks that overlap near 4.5 eV when $V_L = V_R = 0$ are narrow enough to be clearly resolved. They occur at $V = 4.0$ and 4.3 eV. We therefore expect the centers of these peaks occur at $E = 4.0+0.1$ and $4.3+0.1$ eV in the original potential with $V_L = V_R = 0$. Further, these resonances can be unambiguously identified by computing (or probing) the probability distribution function for each at $V_L = V_R = V$, $E = V + \epsilon$.

It is gratifying to know that the individual components of an "unresolvable peak" can be unmasked by "pushing them down" to the energy of a low energy external electron probe beam by varying the boundary conditions.

In Table 19.2 we present the locations of peaks in the transmission probability spectrum $T(E)$ for various combinations of the potential barriers A and B. This includes degeneracies of unresolved/unresolvable peaks. This list does not include

Table 19.1 Energies of peaks in the potential AABB, and source of the resonance in the individual wells AA, AB, and BB

Energy of Peak in AABB		Source of Resonance
0.5		BB
0.7		AB
1.1		AA
2.1		BB
2.9		AB
4.7	Unresolved doublet	AA,BB
6.5		AB
8.3		BB

mirror image barriers (BBA is the mirror image of ABB). The transmission probability spectrum of a barrier is identical to that of its mirror image. This has to do with a symmetry of nature. The symmetry is not a space reflection (or parity) symmetry. Rather, it is the invariance of the Hamiltonian (1.1) under time reversal. To be more specific, if the potential (Hamiltonian) that describes the system is time invariant, the transmission probability is the same whether the potential is probed by particles incident from the left or the right.

Problem. Identify the features in each of the transmission spectra presented in Table 19.2.

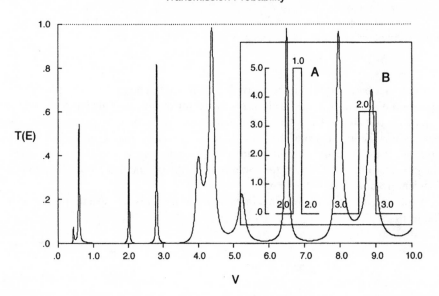

Fig. 19.2 Resolution of the "unresolvable" peaks between 4 and 5 eV in Fig. 19.1. The asymptotic left and right potentials are raised to a potential V, and the resulting potential is probed by an electron beam at energy ϵ ($= 0.1$ eV) above V. The transmission probability is probed as a function of $E = V + \epsilon$. The resolved peaks appear at $V = 4.0$, 4.3 eV, so are close to 4.1 and 4.4 eV in Fig. 19.1.

Table 19.2 Energies at which peaks occur in the transmission probability spectrum $T(E)$ for various combinations of the barriers A and B

Barrier	Peak Energies
AA	1.0, 4.1
AB	0.7, 2.9, 6.6
BB	0.5, 2.1, 4.7, 8.2
AAA	(0.9, 1.2), (3.8, 5.1), 9.3
AAB	0.7, 1.1, 2.9, 4.4, 6.7
ABA	(0.7, 0.8), (2.7, 3.2), (5.9, 7.4)
ABB	0.5, 0.8, 2.1, 2.9, 4.7, 6.5, 8.4
BAB	(0.6, 0.9), (2.6, 3.4), (6.0, 7.1)
BBB	(0.5, 0.6), (2.0, 2.4), (4.3, 5.1), (7.6, 8.9)
AAAA	(0.8, 1.0, 1.4), (3.5, 4.4, 5.5), 8.8
AAAB	0.6, 0.9, 1.3, 2.8, 3.9, 5.1, 6.7, 9.2
AABA	0.7, 0.8, 1.1, 2.6, 3.2, 4.4, 6.0, 7.4, 9.9
AABB	0.5, 0.7, 1.1, 2.1, 2.9, 4.7[a], 6.6, 8.4
ABAB	(0.6, 0.7, 0.9), (2.5, 2.9, 3.5), (5.7, 6.6, 7.7)
ABBA	0.5, 0.8[b], 2.1, 3.0[b], 4.7, 6.3, 6.8, 8.6
BAAB	(0.65, 0.75), 1.2, (2.7, 2.0), 4.5, 6.6, 9.9
ABBB	(0.5, 0.6), 0.8, (2.0, 2.3), 2.9, (4.3, 5.1), 6.5, (7.7, 9.0)
BABB	0.5, 0.6, 0.9, 2.1, 2.6, 3.4, 4.7, 6.0, 7.1, 8.4
BBBB	(0.50, 0.55, 0.62), (1.9, 2.2, 2.5), (4.1, 4.7, 5.3), (7.3, 8.2, 9.2)

Note: Some of the multiplets are grouped in parentheses.
[a] Multiplet unresolvable at any energy resolution.
[b] Multiplet resolvable at finer energy scan.

Part III

Bound States

20

Boundary Conditions

We can learn a great deal about the properties of a potential by firing a beam of electrons at it. Much of the information we can glean is contained in the transmission probability spectrum. Even more information is contained in the phase shift measurements. One objective of such analysis is to determine "what is inside" the potential. That is: Are there bound states? How many? What are their energies? What are their probability distributions? What are their wavefunctions?

Such states are characterized by two closely related properties:

1. Their energies are less than the asymptotic potentials on the left and right:

$$E < V_L, \qquad E < V_R .$$

2. These states are localized or "bound" to the potential. That is, the wavefunction in the asymptotic left- and right-hand regions falls off to zero

$$
\begin{aligned}
\Phi_L &= A_L e^{-\kappa_L x} + B_L e^{+\kappa_L x} \quad \overset{x \to -\infty}{\longrightarrow} \quad 0 \implies A_L = 0 , \\
\Phi_R &= A_R e^{-\kappa_R x} + B_R e^{+\kappa_R x} \quad \overset{x \to +\infty}{\longrightarrow} \quad 0 \implies B_R = 0 .
\end{aligned}
$$

$$(20.1)$$

Here as before $\kappa_L = \sqrt{2m(V_L - E)/\hbar^2}$, $\kappa_R = \sqrt{2m(V_R - E)/\hbar^2}$.

The amplitudes A_L, B_L in the asymptotic left-hand region are related to the amplitudes A_R, B_R in the asymptotic right-hand region by the transfer matrix $T(E)$:

$$
\begin{bmatrix} A \\ B \end{bmatrix}_L = T(E) \begin{bmatrix} A \\ B \end{bmatrix}_R = \begin{bmatrix} t_{11}(E) & t_{12}(E) \\ t_{21}(E) & t_{22}(E) \end{bmatrix} \begin{bmatrix} A \\ B \end{bmatrix}_R .
$$

$$(20.2)$$

In particular, from this equation the boundary conditions (20.1) can be transformed to a condition on the transfer matrix elements as follows:

$$A_L = t_{11}A_R + t_{12}B_R \xrightarrow{(20.1)} t_{11}A_R . \tag{20.3}$$

Since B_R is zero (it is a boundary condition), if A_R is zero, all amplitudes A_j, B_j in the interior of the piecewise constant potential must also be zero since they are linear combinations of A_R and B_R. Therefore, $A_R \neq 0$, and since $A_L = 0$, the boundary condition (20.1) becomes the following condition on the transfer matrix:

$$t_{11}(E) = 0 . \tag{20.4}$$

This condition can be made more explicit by expressing $T(E)$ in terms of the product of transfer matrices $M(V_j; \delta_j)$ for the interior pieces of the potential and the pair of matrices for the two asymptotic regions:

$$\begin{bmatrix} A \\ B \end{bmatrix}_L = \begin{bmatrix} e^{-\kappa_L a_L} & 0 \\ 0 & e^{+\kappa_L a_L} \end{bmatrix}^{-1} \begin{bmatrix} 1 & 1 \\ -\kappa_L & +\kappa_L \end{bmatrix}^{-1} \prod_{j=1}^{N} M(V_j; \delta_j)$$

$$\times \begin{bmatrix} 1 & 1 \\ -\kappa_R & +\kappa_R \end{bmatrix} \begin{bmatrix} e^{-\kappa_R a_R} & 0 \\ 0 & e^{+\kappa_R a_R} \end{bmatrix} \begin{bmatrix} A \\ B \end{bmatrix}_R . \tag{20.5}$$

It is convenient to absorb the exponentials into the definition of the amplitudes. Carrying out the remaining matrix multiplications, we obtain

$$\begin{bmatrix} A \\ B \end{bmatrix}_L' = \begin{bmatrix} \alpha_1 + \alpha_2 & \beta_1 + \beta_2 \\ \beta_1 - \beta_2 & \alpha_1 - \alpha_2 \end{bmatrix} \begin{bmatrix} A \\ B \end{bmatrix}_R' ,$$

$$2\alpha_1 = +m_{11} + \frac{\kappa_R}{\kappa_L}m_{22} , \qquad 2\alpha_2 = -\kappa_R m_{12} - \frac{1}{\kappa_L}m_{21} , \tag{20.6a}$$

$$2\beta_1 = +m_{11} - \frac{\kappa_R}{\kappa_L}m_{22} , \qquad 2\beta_2 = +\kappa_R m_{12} - \frac{1}{\kappa_L}m_{21} , \tag{20.6b}$$

where $A_L' = A_L e^{-\kappa_L a_L}$, and so on, and a_L, a_R are the left- and right-hand boundaries of the potential.

The condition for the existence of bound states is

$$t_{11}(E) = \frac{1}{2}\left(+m_{11} + \frac{\kappa_R}{\kappa_L}m_{22}\right) + \frac{1}{2}\left(-\kappa_R m_{12} - \frac{1}{\kappa_L}m_{21}\right) = 0 . \tag{20.7}$$

The bound states are determined by computing $2t_{11}(E)$ as a function of the energy E and locating the zeroes of this real function. In the often encountered case in which $V_L = V_R$, so that $\kappa_L = \kappa_R = \sqrt{2m(V - E)/\hbar^2} = \kappa$, this simplifies to a search for the zeros of

$$2t_{11}(E) = (m_{11} + m_{22}) - \left(\kappa m_{12} + \frac{m_{21}}{\kappa}\right) . \tag{20.8}$$

21

A Simple Example

To illustrate these results, we compute the bound state energy eigenvalues for the rectangular potential well shown in the inset to Fig. 21.1. The potential in the asymptotic left- and right-hand regions is $V_L = V_R = V$; the potential in the well is $V_1 = 0$ and the width of the well is $\delta_1 = a = 8$ Å. The transfer matrix M for an electron of energy E in the one intermediate region can be read from Table 4.1:

$$M(E) = \begin{bmatrix} \cos ka & -k^{-1} \sin ka \\ k \sin ka & \cos ka \end{bmatrix}, \tag{21.1}$$

where $k = \sqrt{2mE/\hbar^2}$. A simple analytic expression for the energy eigenvalues of the bound states immediately results from (20.8):

$$t_{11}(E) = \cos ka + \frac{1}{2}\left(\frac{\kappa}{k} - \frac{k}{\kappa}\right) \sin ka, \tag{21.2}$$

where $\kappa = \sqrt{2m(V - E)/\hbar^2}$.

Fig. 21.1 presents a plot of $t_{11}(E)$ for an electron in the potential well shown in the inset. The zero crossings of this matrix element occur at energies for which bound state eigenfunctions exist. For the potential shown there are six bound state eigenfunctions.

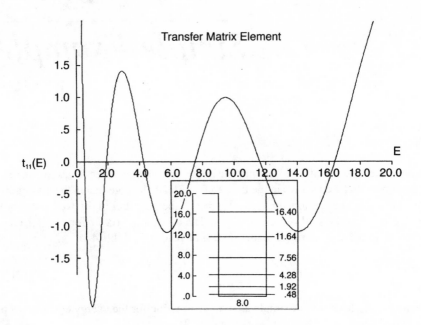

Fig. 21.1 Plot of transfer matrix element $t_{11}(E)$ as a function of energy E for an electron in the potential shown in the inset. Zero crossings of $t_{11}(E)$ define energy eigenvalues. These eigenvalues are shown on the energy scale in the inset.

22
Coding and Validation

Equation (20.7) provides a simple algorithm for computing the bound state energy eigenvalues from the matrix elements of the transfer matrix. However, except for the simplest cases (previous chapter) it is not practical to compute these matrix elements analytically. It is therefore useful to carry out such computations numerically. The algorithm for this numerical computation is straightforward:

- Read in the height V_j and width δ_j of the piecewise constant potential, as well as the asymptotic potential values V_L, V_R on the left and right.

- Choose a value, E, for the incident particle energy ($E < V_L, E < V_R$).

- Compute the real 2×2 matrix $M(V_j; \delta_j)$ for each piece of the potential.

- Multiply these matrices in the order in which they occur, from left to right: $M(E) = \prod_{j=1}^{j=N} M(V_j; \delta_j)$.

- Compute $t_{11}(E)$ in terms of $m_{ij}(E), \kappa_L, \kappa_R$ using (20.7).

In searching for energy eigenvalues by this procedure, this computation must be embedded in a loop that scans over the desired range of energies in which eigenvalues are to be determined.

Once again we emphasize that numerical codes must be validated by extensive testing before one can rely on the results of these computations. One important way to validate a code is to compare its output with results that are analytically available. To this end, we computed the transfer matrix elements $t_{ij}(E)$ for the rectangular well of height V and width L for all energies in the range $0 < E < V$ using the numerical algorithm given above and compared them with the analytically available result given in (21.2). The computation was carried out over a wide range of values of V and L; the comparison was made by subtracting the analytically computed from the numerically computed matrix elements. Nonzero results (to appropriate precision) would invalidate the numerical computation. All differences were zero.

These comparisons for validation purposes are absolutely essential but very uninteresting. These comparisons will not be given here.

Instead, we present some results of these computations that are more interesting from a physical point of view. We computed the energy eigenvalues of an electron in a rectangular well of height $V = 20$ eV and width L, $4 \text{ Å} \leq L \leq 12 \text{ Å}$. The number of eigenvalues depends on L, increasing with L. The narrowest well ($L = 4$ Å) has four eigenvalues. We have plotted E_n, the energy of the nth eigenvalue, as a function of L. The energy dependence increases with n and decreases with L. One could make an inspired guess that $E_n \sim n^2, E_n \sim 1/L^2$. To test how good this inspired guess is, the reduced energy, $E_n / \left[\frac{\hbar^2}{2m} \left(\frac{n\pi}{L} \right)^2 \right]$, is plotted in Fig. 22.1 as a function of L for the four lowest eigenvalues in the potential shown in the inset. These four curves are slowly varying functions of L that asymptotically approach $+1$ as $L \to \infty$. These curves are indistinguishable for L large ($L > 6$ Å) and can only be distinguished for small L. These computations were carried out both analytically and numerically. The results are identical, and only one of these calculations is presented (see Fig. 14.5).

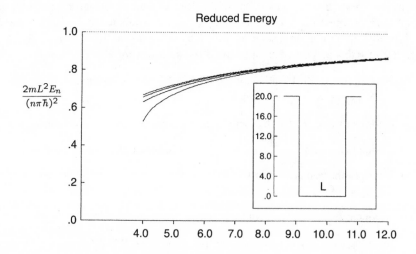

Fig. 22.1   Reduced energy, $E_n / \left(\frac{\hbar^2}{2m} \left(\frac{n\pi}{L} \right)^2 \right)$, plotted as a function of L for the potential shown in the inset. The reduced energy curves for the four lowest levels ($n = 1, 2, 3, 4$, top to bottom, distinguishable at $L = 4$ Å) are slowly varying functions of L that approach $+1$ as $L \to \infty$.

23

Shape of Potential

The results of the energy eigenvalue calculations on the square well potential carried out in chapters 21 and 22 have led to a number of insights. These insights are summarized quantitatively, for the square well potential, in Fig. 22.1. We might expect the *quantitative* results [$E_n \sim \frac{\hbar^2}{2m}\left(\frac{n\pi}{L}\right)^2$] to be specific to the square well potential but the *qualitative* insights to be valid for all potentials. The qualitative insights are

- the number of bound states increases with the depth of the potential

- the number of bound states increases with the width of the potential

- the energy eigenvalues are not sensitive to the shape of the potential, in the sense that small changes in shape produce small displacements of the eigenvalues.

The statements above are vague, as befits qualitative statements. How do we characterize the "shape" of a potential by only two parameters, depth and width? What does a "small change in shape" mean?

We illustrate the meaning of these vague statements by concrete calculations. These will all be carried out on smooth attractive potentials with "Gaussian" shape

$$V(x) = V(1 - e^{-(x/L)^2}) \,. \tag{23.1}$$

For this class of potentials, the parameter V characterizes the depth and L the width of the potential. Potentials with more complicated shapes may be more difficult to characterize, but the qualitative conclusions drawn from these calculations remain unchanged.

In Fig. 23.1 we show the energy eigenvalue spectrum for two different piecewise constant approximations to a Gaussian potential. These approximations are done as previously described (chapter 11, Figs. 11.3, 11.4). Both the five- and eleven-piece approximations to the Gaussian potential have four bound states. The energy eigenvalues differ only slightly in these two approximations. The eigenvalues for the eleven-piece approximation are very close to those of the continuous potential, which are $E = 4.04$, 11.24, 16.68, and 19.80 eV. This calculation shows that the energy eigenvalues change by only small amounts when the shape of the potential undergoes small changes.

Bound State Energies

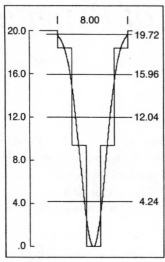 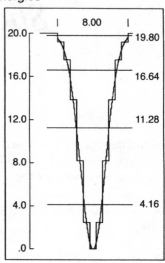

Fig. 23.1 Energy eigenvalues for piecewise constant approximations to the Gaussian potential (23.1), for five and eleven pieces.

Similar calculations confirm that the number of bound states for potentials of type (23.1) increases with the depth, V, of the potential and also with the width, L, of the potential.

It would be very useful if we could give answers to the following questions: For an arbitrary binding potential $V(x)$

- How many eigenstates does $V(x)$ support?

- How many bound states are in $V(x)$ below an energy E?

At the present time we can give only an approximate answer to these questions. In chapter 36 we will provide a much better method to answer these two questions. However, the results presented below are not substantially different.

Since we address these questions later, we only discuss the most important question at present. It is not unreasonable to assume that a bound state will occur when the total phase shift made by an electron in a round-trip circuit is an integer multiple of 2π. This is Bohr's quantization condition. If the electron is confined to the region of space $a \leq x \leq b$, where $E - V(x) \geq 0$ (classically allowed region), then the total phase shift in going from a to b and back again consists of four terms

$$\int_a^b \frac{p(x)}{\hbar} \, dx + \Delta\phi_R + \int_b^a \frac{p(x)}{\hbar} \, dx + \Delta\phi_L = 2\pi n \,. \tag{23.2}$$

The phase shifts in propagating from a to b and from b to a are equal. The phase shift $\Delta\phi_R$ incurred on reflection from the right-hand edge of the potential at $x = b$ ranges from 0 when $V_R - E \simeq 0$ to π when the particle is deep in a steep well. Similar remarks hold for the phase shift $\Delta\phi_L$ incurred in reflection at $x = a$. This uncertainty in the phase shifts on reflection produces a slight uncertainty in our estimate for n:

$$n \sim \frac{2}{2\pi} \int_a^b \sqrt{\frac{2m}{\hbar^2}(E - V(x))} \, dx \,. \tag{23.3}$$

In Fig. 23.2 we plot the number of bound states, n, as a function of the action, $\oint pdq/2\pi\hbar \sim \int_a^b pdq/\pi\hbar$, for the Gaussian potential. In this computation we have kept the depth and width of the potential constant and varied the action by varying the energy E. The general results, however, are independent of what is held fixed and what is allowed to vary. In fact, the results are independent of the shape of the potential.

These results allow us to conclude that the number of bound states that a potential can support, or the number of bound states with energy less than E, is not determined by vaguely defined parameters V, L that separately characterize the depth and width of a potential, but rather by a well-defined parameter, the action $A = \oint p \, dq \sim 2 \int_a^b p \, dq$, by means of the dimensionless ratio $n \sim A/(2\pi\hbar) = A/h$.

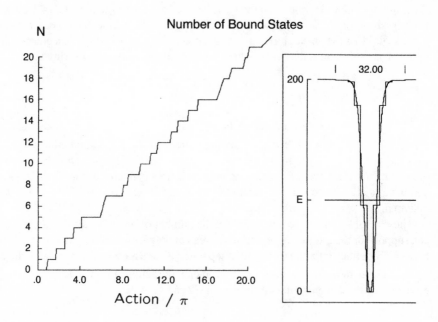

Fig. 23.2 Number of bound states with energy less than E as a function of $\int_a^b p\,dx/\pi\hbar$, for the potential (23.1) with $V = 200.0$ eV, $L = 4.0$ Å. The integral extends over the classically allowed region.

24

Dependence on Parameters

We now study in more detail how the energy eigenvalues depend on the parameters that characterize a potential. For this study we use the simple square well of depth V and width L. We do this knowing that the quantitative results of these calculations will depend on the shape of the potential chosen but the qualitative results do not.

In Figs. 24.1 and 24.2 we plot the energy eigenvalues of the square well potential obtained when one of the two shape parameters is varied and the other is held constant.

In Fig. 24.1 we fix $L = 8.0$ Å and plot the energy eigenvalues in the range $0 < E < V$ for V in the range $0 < V \leq 20.0$ eV. Bound states exist only below the diagonal line in the plot. Several important features of this figure deserve mention:

- There is always one bound state, no matter how shallow the potential. This is true for any one-dimensional potential.

- As V increases new bound states come into existence at the top of the potential ($E = V$).

- As the potential height increases, the energy of the bound states also increases.

- The bound state energies approach the limit $E_n = \frac{\hbar^2}{2m} \left(\frac{n\pi}{L} \right)^2$ asymptotically from below as $V \rightarrow \infty$.

In Fig. 24.2 we fix $V = 20.0$ eV and plot the energy eigenvalues in the range $0 < E < V = 20.0$ eV for L in the range 4.0 Å $\leq L \leq 12.0$ Å. Several important features of this figure also deserve mention:

- As L increases new bound states come into existence at the top of the well ($E = V$).

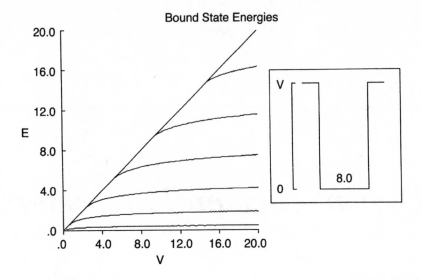

Fig. 24.1 Bound state energies for square well potential (inset) as function of height V. New bound states are created at the top of the well at $E = V$ as V increases.

- Bound states decrease in energy as L increases, approaching the limit $E_n = \frac{\hbar^2}{2m}\left(\frac{n\pi}{L}\right)^2$ from below as $L \to \infty$ (see Fig. 22.1).

From these discussions we can draw the following conclusions, which are valid for all potentials

- When new bound states are created due to a change in the shape of a potential, they occur at the "top" of the potential $E = V_L = V_R$.

- When bound states are destroyed by changing the shape of a potential (e.g., decreasing V or L), they disappear from the top of the potential.

At what energies do new bound states appear in the square well potential? A simple way to determine these energies is illustrated in Fig. 24.3. This figure provides an expansion of the range from 5.0 to 10.0 eV shown in Fig. 24.1. In this range there are only two bound states. We will determine the energy at which the lower of these two states ($n = 4$) is created. To do this, we compute the energy of this state for some value of the potential V (9.9 eV in this case). Chose this energy as the new value of the potential (follow paths 1 and 2). The bound state will exist for this new value of V, since the curve E_n is monotonic. By repeating this process we can locate the energies at which these states come into existence. We find by this process and by methods to be described in the following chapter that the nth eigenstate is created at potential values

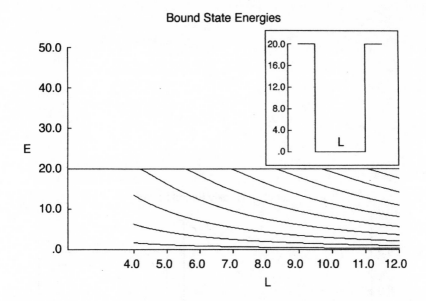

Fig. 24.2 Bound state energies for square well potential (inset) as function of width L. New bound states are created at the top of the well at $E = V$ as L increases.

$$V_n = E_n = \frac{\hbar^2}{2m} \left[\frac{(n-1)\pi}{L} \right]^2 . \tag{24.1}$$

Problem. Show that the $E_n(V)$ versus V curves have quadratic tangency with the diagonal $E = V$ in Figs. 24.1 and 24.3.

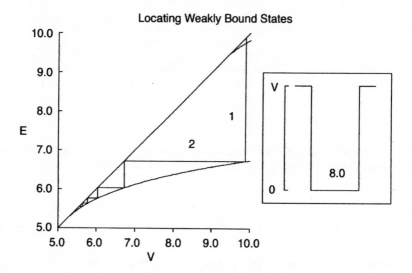

Fig. 24.3   Locating energies at which new bound states are created. Bound state $n = 4$ exists at $V = 9.9$ eV with energy $E = 6.8$ eV. Set $V = 6.8$ eV and find new energy of this state. Iterate to convergence, to find $E_n = \frac{\hbar^2}{2m} \left[\frac{(n-1)\pi}{L} \right]^2 = 5.28$ eV.

25

Relation between Bound and Scattering States

Figs. 24.1 and 24.2 are instructive yet somehow unsatisfying. They suggest that new bound states appear "out of thin air." Nothing appears out of thin air. To better appreciate these figures, it is useful to fill in some details in the $E > V$ half of these figures.

The most prominent features in the transmission probability spectrum $T(E)$ are the peaks. For a square well potential of width L, the locations of the peaks has already been computed in equation (9.6). They occur at energy

$$E_n = \frac{\hbar^2}{2m} \left(\frac{n\pi}{L}\right)^2, \qquad E > V \tag{25.1}$$

above the bottom of the potential. In Figs. 25.1 and 25.2 we reproduce Figs. 24.1 and 24.2, which give the bound state energies of the square well potential and include the location (dotted curves) of the peaks in the transmission probability spectrum. It is apparent from these figures that the bound states in fact do not simply appear "out of thin air." As the potential well is deepened (Fig. 25.1) or widened (Fig. 25.2), a transmission resonance approaches the top of the well (i.e., $V \uparrow E$). As it does so, its half width becomes increasingly small and the resonance becomes increasingly narrow. When V increases above E, the resonance is transformed into a bound state.

In Fig. 25.3 we have redrawn Fig. 25.1 with a change in scale. The scales are now the square root of the potential depth, $\sqrt{V}$, and the square root of the energy at which transmission resonances or bound states occur, $\sqrt{E}$. Resonance peaks occur above the diagonal $\sqrt{E} = \sqrt{V}$, bound states occur below. This figure makes clear that transmission resonance peaks have fixed energy $E_n = \frac{\hbar^2}{2m} \left(\frac{n\pi}{L}\right)^2$.

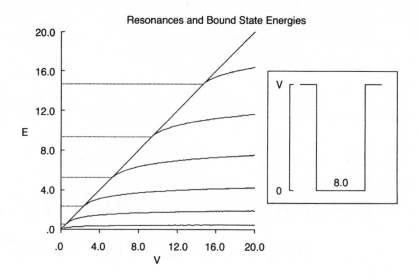

Fig. 25.1 Peaks in the transmission probability spectrum (dotted, $E > V$) and bound state energy spectrum (solid, $E < V$) as a function of potential height V for the square well potential (inset, $L = 8.0$ Å). Peaks and bound states appear to be continuations of each other.

In order to see whether these phenomena are typical for all binding potentials or specific to the square well potential, we have carried out a similar series of calculations for the Gaussian potential (23.1). For these calculations we have chosen a five-piece approximation to this potential, fixed $L = 2$ Å and varied V from 0 to 20.0 eV. In Fig. 25.4 we plot the location of the transmission probability peaks (dotted, $E > V$) and the bound state energy eigenvalues ($E < V$, solid) as a function of V. This plot clearly shows that each bound state is formed by passage of a transmission resonance through the $V = E$ barrier. A rescaled plot ($\sqrt{E}$ versus $\sqrt{V}$; see Fig. 25.3) of the same data is presented in Fig. 25.5.

While Figs. 25.1 and 25.3 seem to suggest that there is a one-to-one correspondence between bound states ($E < V$) and transmission resonances ($E > V$), Figs. 25.4 and 25.5 suggest otherwise. The latter figures show strange "bifurcation phenomena" among the peaks. For example, there are several instances in which a single peak appears to split into two (we have already explored this phenomenon in chapter 19).

In order to understand this phenomenon better, assume a transmission probability spectrum consists of a superposition of two Lorentzian peaks,

$$L_i(E) = \frac{A_i}{1 + [(E - E_i) / (x\Delta E_i)]^2} \,, \qquad (25.2)$$

centered at $E = E_i$ with half width $x\Delta E_i$. As the peaks at E_i approach the top of the potential, they become narrower and narrower. This is represented by $x \rightarrow 0$ in

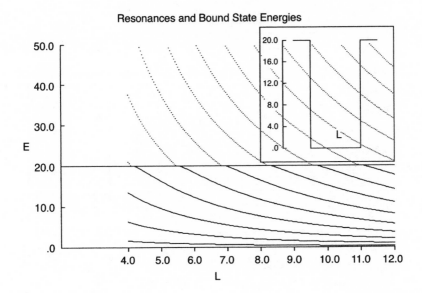

Fig. 25.2 Peaks in the transmission probability spectrum (dotted, $E > V$) and bound state energy eigenvalue spectrum (solid, $E < V$) as a function of width L for the square well potential (inset, $V = 20.0$ eV).

the functions above. If the peaks are too broad, they cannot be resolved. As the half widths shrink, they become resolvable. Fig. 25.6 provides a plot of the location of the peak(s) of the transmission intensity

$$T(E) = L_1(E) + L_2(E) \tag{25.3}$$

as a function of the parameter $1 - x$. For $x = 1$ the peaks overlap and cannot be resolved, so $T(E)$ has a single maximum. As x decreases, the half width narrows, and a new maximum and minimum appear on the shoulder of the original peak. As x decreases further, the new local maximum and minimum move apart, and the minimum "pushes" the original maximum away. As $x \to 0$ the two peaks are clearly resolved. The maxima in Fig. 25.6 are drawn with a dark line. The minimum, which appears between the two maxima, is drawn with a lighter line.

The bifurcations that appear in Figs. 25.4 and 25.5 occur because the half widths of overlapping peaks narrow sufficiently to resolve the separate peaks as the threshold $V = E$ approaches.

Figs. 25.1–25.5 reveal clearly that each bound state is created as a transmission resonance sinks below the top of the potential. While Figs. 25.1–25.3 clearly suggest a one-to-one correspondence between peaks and resonances, Figs. 25.4 and 25.5 show more bound states than resonances simply because broader resonances ($E \gg V$) overlap and cannot be resolved. Thus, there is in fact a one-to-one correspondence

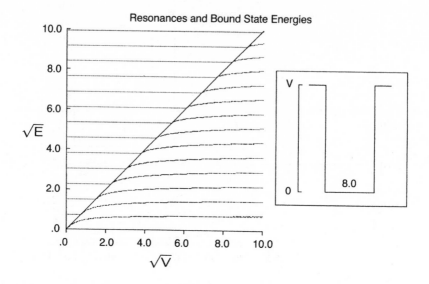

Fig. 25.3 Presentation of the data in Fig. 25.1 $(0 < E, V < 100.0 \text{ eV})$ using rescaled axes. The dotted horizontal lines (peaks) are equally spaced in this representation, and their extensions are asymptotes of the rising energy eigenvalue curves.

between bound states and resonances, but some of the resonances may not be re-solvable, and so appear as a single peak in plots such as Figs. 25.4 and 25.5 until $V \simeq E$.

These statements are true for all binding potentials. However, they are qualitative. We now wish to make them quantitative. To do this, we express the transfer matrix element $t_{11}(E)$ in terms of the matrix elements $m_{ij}(E)$ of the product of transfer matrices for the piecewise constant parts of the potential with $V_L = V_R$

$$E > V \quad t_{11}(E) = \frac{1}{2}(m_{11} + m_{22}) + \frac{i}{2}\left(m_{12}k - \frac{m_{21}}{k}\right) \xrightarrow{E \downarrow V} -\frac{im_{21}}{2k},$$

$$E < V \quad t_{11}(E) = \frac{1}{2}(m_{11} + m_{22}) - \frac{1}{2}\left(m_{12}\kappa + \frac{m_{21}}{\kappa}\right) \xrightarrow{E \uparrow V} -\frac{m_{21}}{2\kappa}.$$

$$(25.4)$$

The matrix elements $m_{ij}(E)$ are smooth functions of the energy E and the param-eters $(V_j; \delta_j)$ that define the shape of the potential. These matrix elements have no singularities. The singularities of $t_{11}(E)$ occur only in the factors $1/k$ and $1/\kappa$. As E approaches V from above or below, $t_{11}(E) \sim m_{21}/\sqrt{|E - V|}$. Since $t_{11}(E)$ blows up if $m_{21} \neq 0$, $T(E) \sim |1/t_{11}(E)|^2$ becomes very small. A transmission resonance can occur as $k \to 0$ only if $m_{21}(E = V) = 0$. Similarly, a bound state energy eigenvalue can occur as $\kappa \to 0$ only if $m_{21}(E = V) = 0$. The condition

$$m_{21}(E = V) = 0 \qquad (25.5)$$

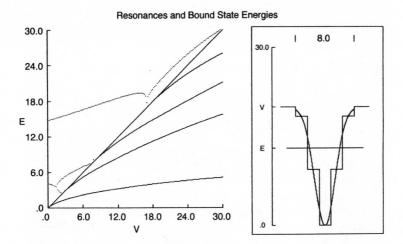

Fig. 25.4 Peaks in the transmission probability spectrum (dotted, $E > V$) and bound state energy spectrum (solid, $E < V$) as a function of potential depth V for the five-piece approximation to the Gaussian potential with $L = 2.0$ Å (inset). Peaks show bifurcations.

provides a quantitative description of the conversion of a peak in the transmission probability spectrum to a bound state energy eigenvalue as E passes from above to below the asymptotic potential values $V_L = V_R$.

To illustrate this quantitative statement, the matrix element $m_{21}(E)$ for the square well potential is

$$m_{21}(E) = k \, \sin kL \,. \tag{25.6}$$

Therefore the energies at which transmission resonances become bound states are defined by $kL = n\pi$, or

$$E_n = \frac{\hbar^2}{2m} \left(\frac{n\pi}{L} \right)^2 \,. \tag{25.7}$$

Problem. For what values of the depth V ($0 < V < 20.0$ eV) do new bound states appear for the eleven-piece approximation to the Gaussian potential?

Solution. Fix V. Then set $E = V$ and compute the transfer matrix $M(E) = \prod_{j=1}^{N} M(V_j; \delta_j)$. Plot the matrix element $m_{21}(E)$ as a function of V. This plot is presented in Fig. 25.7. The zero crossings occur at $V = 0.0$, 2.56, 8.28, and 17.10 eV. The zero at $E = 0.0$ eV is simply a reflection of the fact that an arbitrarily weak potential creates a bound state in one dimension. For V slightly below 17.10 eV, a transmission resonance peak occurs just above the asymptotic limits. For V slightly above 17.10 eV, a bound state exists just below the asymptotic limits.

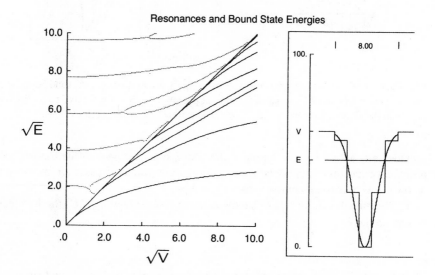

Fig. 25.5 Presentation of the data in Fig. 25.4 ($0 < V, E < 100.0$ eV) using rescaled axes. The bifurcations show an alternating structure.

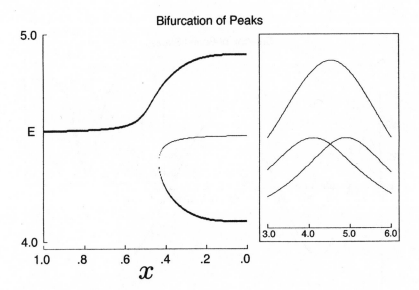

Fig. 25.6 Location of peaks (dark curves) for two overlapping Lorentzian curves $L_i(E) = A_i / \{1 + [(E - E_i)/(x\Delta E_i)]^2\}$, as a function of line width. The new peak and minimum (lighter curve) appear in a saddle node bifurcation. Here $A_1 = A_2 = 0.8, E_1 = 4.1, E_2 = 4.9, \Delta E_1 = 1.5, \Delta E_2 = 1.4$ eV. The two individual peaks, and their sum, are shown at the right ($x = 1$).

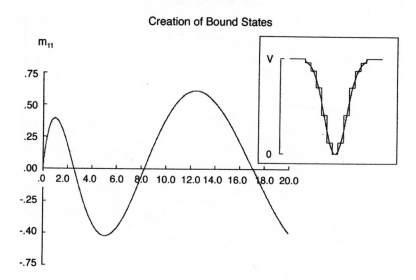

Fig. 25.7 Zeros of $m_{21}(E = V)$ for the Gaussian potential (eleven-piece approximation) shown in inset. The zeros identify energies at which a transmission peak becomes an energy eigenvalue.

26

Double and Multiple Well Potentials

Up to this point we have studied the properties of potential wells that have one minimum. What can we expect if the potential well has two or more minima separated by intermediate maxima? Such potentials could consist of two square well potentials separated by an intermediate barrier or two Gaussian potentials placed side by side.

If the potential wells are different and separated by a large distance, one could expect the spectrum of bound state energies would be easy to predict. The spectrum should consist of the bound state energies for the two individual wells. This intuitive guess is correct, and we will return to this case subsequently to verify it.

What happens if the wells are identical? We might expect each bound state to occur twice, with identical energies. We have already done analogous calculations on multiple barrier potentials, and on the basis of these calculations, we could anticipate a slightly different result. Two barriers contain a single well between them, and the transmission probability spectrum shows a series of single peaks in the classically forbidden regime (see Fig. 14.1). Three identical, equally spaced barriers contain two identical wells between them, and the transmission probability spectrum contains a series of split peaks in the classically forbidden regime (see Fig. 15.1).

By analogy and intuition, we might expect that the spectrum of two identical potentials would consist of doublets, with each bound state energy of the single well split into a doublet in the double well potential. Fig. 26.1 presents results to confirm this inspired guess. The energy eigenvalues for a single square well potential of depth 5.0 eV and width 6.0 Å are 0.62, 2.39, and 4.77 eV. This is shown on the left in Fig. 26.1. Two identical wells, separated by a barrier of height 5.0 eV ($= V_L = V_R$) and width 2.0 Å support five bound states. This is shown on the right in Fig. 26.1. The pair at 0.59 and 0.65 eV are the doublet arising from the state at 0.62 eV in the single-well potential. These two states are 0.03 eV below and above the original singlet at 0.62 eV. The doublet with $E = 2.26$ and 2.54 eV in the double well potential arise from the isolated level at 2.39 in the single-well potential. These two states are repelled 0.13 eV (below) and 0.15 eV (above) the position of the singlet.

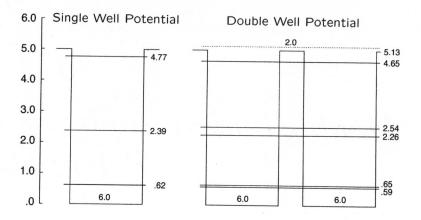

Fig. 26.1   Spectra of single- and double-well potentials. Each isolated state in the single well splits into a doublet in the two well potential. One of the states in the upper doublet has been repelled above the ionization threshold. It is identified as a peak in the transmission probability spectrum.

But where is the doublet associated with the isolated level at 4.77 eV in the single-well potential? One bound state at 4.65 is present in the double well potential. It is 0.12 eV below the position of the singlet. We should expect the other member of this doublet to be somewhat more than 0.12 eV above the position of the singlet. It is, but it is enough above the singlet at 4.77 eV that it is not a bound state. It is a resonant scattering state, identified by a peak at 5.13 eV in the transmission probability spectrum. It is indicated by a dotted line in the energy-level diagram. The doublet in the double well potential that is associated with the very weakly bound singlet at 4.77 eV (it is only 0.23 eV below the "ionization threshold") actually consists of a bound state at 4.65 eV and an unbound state (transmission resonance) at 5.13 eV. These three doublets are split by 0.06, 0.28, and 0.48 eV.

Potentials consisting of equally spaced multiple identical wells behave similarly. In Fig. 26.2 we show the spectrum arising from three and four wells identical to that shown in Fig. 26.1. In the three well potential (top) the ground state splits into a triplet which is unresolved at the energy resolution shown. The state at 2.39 eV splits into a triplet of unequally spaced levels. The loosely bound state at 4.77 eV splits into a triplet consisting of two bound states ($E < 5.0$ eV) and one resonant scattering state ($E > 5.0$ eV).

The four well potential (bottom, Fig. 26.2) behaves similarly. The lowest-lying state splits into an unresolved 4-plet at 0.62 eV. The middle level splits into a 4-plet

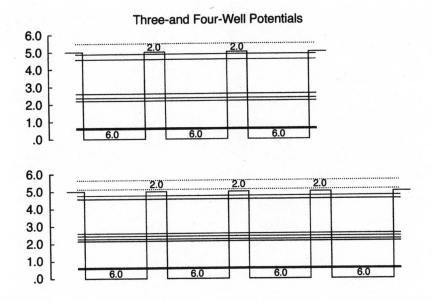

Fig. 26.2 Energy-level spectrum for three and four identical wells. Isolated states in the single well split into multiplets in multiple well potentials. Some states (dotted) may be repelled above the ionization threshold.

with center of gravity near but slightly above 2.39 eV. The loosely bound state at 4.77 eV splits into a quartet. Two of these states are bound and two unbound. In these energy-level diagrams, the bound states are shown as solid lines, the unbound states as dotted lines.

The behavior of N-identical equally spaced wells can be easily extrapolated from this discussion and Figs. 26.1 and 26.2.

The phenomena observed for the transmission spectrum of $N+1$ identical, equally spaced barriers (forming N identical equally spaced wells) described in Part II, chapter 15, and the "bound" state spectrum of N identical equally spaced wells described here are very closely related. We defer a discussion of the width of an N-tuplet of bound states, just as we deferred a discussion of the width of an N-tuplet of transmission resonance peaks. The width of a band of states, either bound or resonant, will be determined in Part IV, chapter 40.

27

Level Splitting

We return now to a question left unanswered in the previous chapter. What determines the splitting between almost degenerate levels in a double well potential?

To answer this question, we could take the approach adopted in Part II, chapter 12 and presented in Figs. 12.1–12.3. If we study the splitting, ΔE, of a given doublet as a function of the separation, D, between wells, we find by plots analogous to Fig. 12.1 that $\Delta E \sim e^{-\gamma_1 D}$. If we study the splitting as a function of the depth below the top of the barrier separating the two wells, we find by plots analogous to Fig. 12.2 that $\Delta E \sim e^{-\gamma_2 \sqrt{V - E_n}}$, where E_n is the energy of the corresponding singlet in the single well potential. We have seen this kind of behavior before (Part II, chapter 12) and can therefore make an inspired guess that the splitting falls off exponentially with the action, or more accurately, with its analytic continuation

$$\ln \Delta E \sim -\int_a^b \kappa(x)dx = -\int_a^b \sqrt{\frac{2m}{\hbar^2}(V(x) - E_n)} \, dx \, . \qquad (27.1)$$

It is therefore useful to compute the splitting of doublets and plot this splitting as a function of action.

This is done in Fig. 27.1. We have chosen a potential well that supports three bound states. The single well has $V_R = 4.0$ eV, $V_L = H \geq V_R, L = 8.0$ Å. When placed back to back with a mirror image of itself, a double well is formed with $V_L = V_R = 4.0$ eV. Both wells have length 8.0 Å. They are separated by an intermediate barrier whose height, H, and width, D, are varied. For the range of parameters studied ($L = 8.0$ Å, $V_L = V_R = 4.0\text{Å} \leq H \leq 9.0\text{Å}$), the single well has three bound states. The doublets for the double well potential are shown in the inset to Fig. 27.1. We have computed the splitting between doublets and plotted the negative logarithm against

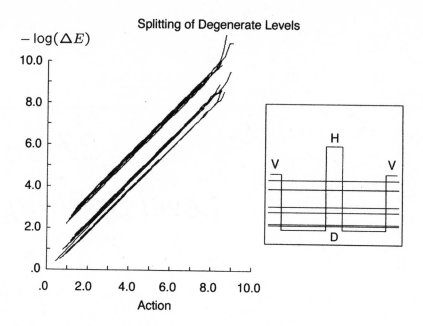

Fig. 27.1 Logarithm of level splitting as a function of "Action," or tunneling probability amplitude. For this potential $V_L = V_R = 4.0$ eV, and each well is 8.0 Å wide. The height H and width D of the intermediate barrier are varied to change the level splitting and action: 4.0 eV $\leq H \leq$ 9.0 eV, 1.0 Å $\leq D \leq$ 8.0 Å.

the action $\int_a^b \sqrt{2m(V(x) - E_n)/\hbar^2} \, dx$, where E_n is the energy of the parent state in the single well potential. This plot, shown in Fig. 27.1, consists of three sets of straight lines, corresponding to the three different levels. The upper set of six lines corresponds to H values of 4.0, 5.0, $\cdots$, 9.0 eV, with D scanned slowly from 1.0 Å to 8.0 Å for the lowest-lying doublet. The next two sets of six straight lines are for the middle and upper doublets.

From these calculations we conclude that the splitting between levels that would otherwise be degenerate with energy E_n falls off exponentially with the action: $\Delta E \sim \exp\left[-\int_a^b \sqrt{2m(V(x) - E_n)/\hbar^2} dx\right]$. In essence, the splitting is related to the probability amplitude for tunneling through the barrier separating the two wells (see equation (12.1)).

28

Symmetry Breaking

When two potential wells are separated by a barrier, we would expect that the energy eigenvalues of the combined potentials are simply the eigenvalues of the two isolated potentials. If two eigenvalues are degenerate, either by accident or by symmetry, then these degenerate levels will split, and the magnitude of this splitting can be estimated by (27.1).

In general, accidents don't happen. If two wells are different, their energy eigenvalues will be different, no degeneracies will occur, no splitting will take place. However, if we deform one well or both wells, then sooner or later a level in one well will become equal to a level in the other. Will these levels become degenerate in the combined potential? We expect not. How will these levels fail to become degenerate?

To answer such questions we have placed two different wells adjacent to each other. These wells are shown in the inset to Fig. 28.1. The shape of these wells is varied by changing their depth. In particular, both wells have length $L = 6.0$ Å and are separated by a barrier with $V = 0.0$ eV and width 2.0 Å. The depth of the wells is varied by adjusting a parameter V, representing an externally applied voltage. One well has depth $-5.0 - V$ eV, the other $-5.0 + V$ eV. When $V = 0$ the wells are identical.

We have computed the spectrum of this double well system as the parameter V is varied. Since bound and scattering states differ only by accident of boundary conditions, we search for bound states and resonances in the range $-5.0 - |V| \leq E \leq +10.0$ eV. Bound states are shown as solid curves; unbound states are dotted.

A number of observations are in order:

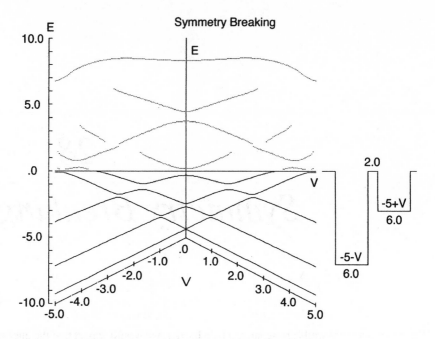

Fig. 28.1 Avoided crossings among bound states and resonances in the two well potential shown. The symmetry breaking parameter is V and the plot is symmetric under $V \rightarrow -V$.

- Bound state eigenvalues do not cross. Two avoided crossings can be seen on the symmetry axis $V = 0$ at $E \sim -4.5$ and ~ -2.8 eV. Other avoided crossings occur at $V \sim \pm 1.6$ eV, $E \sim -1.0$ eV; and $V \sim \pm 2.5$ eV, $E \sim -2$ eV.

- Bound state eigenvalues show avoided crossings with transmission peak resonances. One avoided crossing can be seen at $V \sim \pm 4.9$ eV, $E \sim 0$ eV.

- Resonances exhibit avoided crossings. In fact, they exhibit avoided crossings by three mechanisms:

  1. If the wells are not identical, the transmission peaks are not identical. When they approach too closely they become unresolvable. One seems to disappear. In fact, it disappears in a saddle node bifurcation with the minimum that separates it from the nearby peak (see Figs. 25.4–25.6). This phenomenon is seen at $V \sim \pm 2.5$ eV, $E \sim 2$ eV as the peak with rising energy ($V > 0$) is swallowed at $V \sim 1.8$ eV by the other with higher energy and then reappears at $V \sim 3.0$ eV. The "crossing" took place hidden from view.

2. If the wells are not identical but the peaks meet at such low energy that their half widths are smaller than the separation between their centers, then the peaks are always resolvable and undergo an avoided crossing. This occurs at $V \sim \pm 4$ eV and $E \sim 0.5$ eV.

3. If the wells are identical, the peaks are split and resolvable. This occurs at $V = 0$ and $E \sim 4.5$ eV.

- Some states are intermittently bound states and resonances. The most weakly bound state at $V = 0$ remains bound for $|V| < 3.5$ eV but is a resonance for $|V| > 3.5$ eV.

As V increases from -5.0 eV to $+5.0$ eV, some of the energies increase while others decrease. Those which increase belong to the right-hand well. For $V = -5.0$ eV, the left-hand well ceases to exist and the right-hand well has depth 10.0 eV. This well supports bound states with energies about 0.7, 2.8, and 6.2 eV above the bottom of the well. It also supports resonances about 10 eV and 16.5 eV above the bottom of the well. As V increases, the energy of these states and resonances remains almost unchanged relative to the bottom of the well. Similar remarks hold for the left-hand well at $V = +5.0$ eV. As V varies, these levels are bound to run into each other. We have seen that they avoid crossing each other whether they are bound states or resonances.

We can break the symmetry of the two identical wells of depth 5.0 eV and width 6.0 Å, not by varying their depths, as in Fig. 28.1, but by varying their widths, as shown in Fig. 28.2. We see the same phenomena of avoided crossings in this mode of symmetry breaking as seen previously, when the well depth was varied.

We present a similar calculation for deeper, narrower wells in Fig. 28.3, in which these two wells have widths $5 \pm D$ Å. We recall that the eigenvalues and resonances of a square well behave like $E_n \sim \frac{\hbar^2}{2m} \left(\frac{n\pi}{L} \right)^2$. The energies which increase like $(5-D)^{-2}$ to the right belong to the right-hand well. Those increasing like $(5+D)^{-2}$ to the left belong to the left-hand well. As D is varied, these levels vary in their standard way $(5 \pm D)^{-2}$ except when they are perturbed by a nearby level. Rather than cross, the states avoid crossing by repelling each other, as described above.

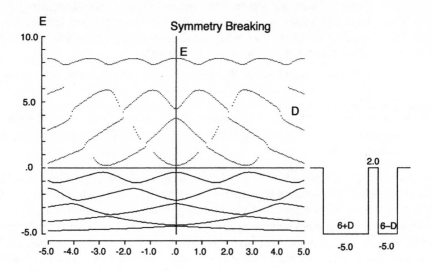

Fig. 28.2 Avoided crossings among bound states and resonances in the two well potential shown. The symmetry breaking parameter is D and the plot is symmetric under $D \to -D$.

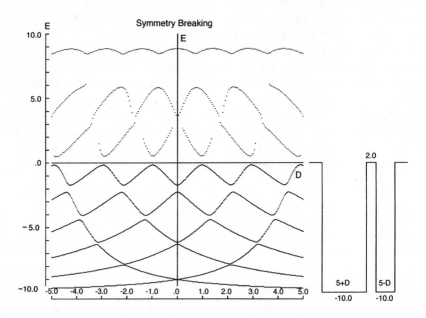

Fig. 28.3 Eigenvalues and resonances for the two well potential shown. These energies behave like $E_n^{\pm} = \frac{\hbar^2}{2m} \left(\frac{n\pi}{5 \pm D} \right)^2$ except where two energies become degenerate.

29

Wavefunctions

In order to identify energy eigenvalues in multiple well potentials, it is useful to compute the corresponding eigenfunctions, or wavefunctions.

We have solved this problem in Part I, chapter 3 and again in Part II, chapter 16. In Part I we used as basis functions for solutions to Schrödinger's equation traveling waves $e^{\pm ikx}$ $(E > V)$ and decaying/growing solutions $e^{\mp \kappa x}$ $(E < V)$. In Part II, chapter 16, we have used as a basis set the trigonometric functions $\cos kx$, $\sin kx$ $(E > V)$ and hyperbolic functions $\cosh \kappa x$, $\sinh \kappa x$ $(E < V)$. The choice of basis is dictated simply by convenience.

In either case we apply boundary conditions to impose conditions on the amplitudes $\begin{bmatrix} A \\ B \end{bmatrix}_{0 \ or \ L}$, $\begin{bmatrix} A \\ B \end{bmatrix}_{N+1 \ or \ R}$ in the asymptotic left- and/or right-hand regions. Then transfer matrices are constructed to relate the amplitudes $\begin{bmatrix} A \\ B \end{bmatrix}_j$ in region j to amplitudes $\begin{bmatrix} A \\ B \end{bmatrix}_{j+1}$ in region $j + 1$.

To construct bound state wavefunctions, it is convenient to use trigonometric / hyperbolic cosines and sines rather than the exponential functions. The boundary condition in the right-hand region is

$$
\begin{aligned}
\Phi_{N+1} &= A_{N+1} \cosh \kappa x + B_{N+1} \sinh \kappa x \\
&= \tfrac{1}{2}\left(A_{N+1} + B_{N+1}\right) e^{+\kappa x} + \tfrac{1}{2}\left(A_{N+1} - B_{N+1}\right) e^{-\kappa x} \xrightarrow{x \to +\infty} 0 .
\end{aligned}
\tag{29.1}
$$

This requires $A_{N+1} + B_{N+1} = 0$. Any nonzero choice $B_{N+1} = -A_{N+1}$ is suitable for initializing the computation of all amplitudes $\begin{bmatrix} A \\ B \end{bmatrix}_j$. However, if we choose A_{N+1} to be real, then so is B_{N+1}. Since all transfer matrices between adjacent regions are also real (see equation (16.2)), all amplitudes $\begin{bmatrix} A \\ B \end{bmatrix}_j , j = 0, 1, \cdots, N, N+1$

are also real. Since the basis functions in each region are also real, we conclude that *in one dimension every bound state eigenfunction can be made real.*

In initializing the computation of the amplitudes $\begin{bmatrix} A \\ B \end{bmatrix}_{N+1}$, it doesn't matter whether we choose the nonzero amplitude $A_{N+1} = -B_{N+1} = 1, 10, 0.1, -10^{-2}$, $1 - i$, and so on. All wavefunctions so constructed will be proportional to each other. However, it is useful to present a wavefunction in a standard or normalized form.

The wavefunction represents a probability amplitude and its square represents a probability density. A bound state is localized or literally "bound" to its potential. We therefore normalize bound state wavefunctions by the condition

$$\int_{-\infty}^{+\infty} \overline{\Phi(x)}\Phi(x)dx = \int_{-\infty}^{+\infty} |\Phi(x)|^2 \, dx = +1 . \tag{29.2}$$

All wavefunctions described in this chapter are normalized by this condition. Normalization is implemented numerically by breaking the integral into three parts:

$$\int_{-\infty}^{+\infty} |\Phi(x)|^2 \, dx = \int_{-\infty}^{a_0} |\Phi(x)|^2 \, dx + \int_{a_0}^{a_N} |\Phi(x)|^2 \, dx + \int_{a_N}^{+\infty} |\Phi(x)|^2 \, dx = +1 . \tag{29.3}$$

The integral from $-\infty$ to the beginning of the potential at a_0 is easily computed analytically. It is $(A_0 + B_0)^2 e^{2\kappa_L a_0}/8\kappa_L$. Similarly, the integral from the end of the potential at a_N to $+\infty$ is $(A_{N+1} - B_{N+1})^2 e^{-2\kappa_R a_N}/8\kappa_R$. The remainder of the integral is the sum of contributions from each of the N pieces of the potential:

$$\int_{a_0}^{a_N} |\Phi(x)|^2 \, dx = \sum_{j=1}^{N} \int_{a_{j-1}}^{a_j} |A_j \cos(\text{h})kx + B_j \sin(\text{h})kx|^2 \, dx . \tag{29.4}$$

Each of these terms is easily evaluated.

In Fig. 29.1 we plot eigenfunctions for each of the bound states supported by three different potentials. These potentials are the square well with depth 20.0 eV and width 6.0 Å; the fifty-one-piece approximation to the Gaussian potential $V(x) = V_0[1 - e^{-(x/L)^2}]$, $V_0 = 20.0$ eV and $L = 2.0$ Å; and a two-step potential with constant parts at 0.0 and 7.0 eV, each of width 3.0 Å, with $V_L = V_R = 20.0$ eV. For each potential the energy eigenvalues were first determined by locating the zeros of $t_{11}(E)$ accurately. These energies were then used to compute the amplitudes $\begin{bmatrix} A \\ B \end{bmatrix}_j$ following the algorithm described above. Finally, all wavefunctions were normalized to unity and plotted to the same scale on a baseline that locates the energy eigenvalue.

The square well potential supports five bound states, the other two support four. We observe that these wavefunctions are in some sense very similar. In each case, the lowest-energy wavefunction has no nodes (zero crossings). The number of nodes in

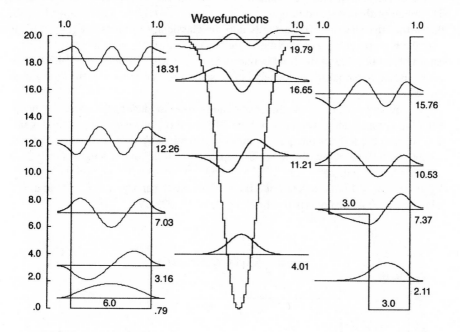

Fig. 29.1 Wavefunctions for three potentials. The energy E_n of the wavefunction $\psi_n(x)$ increases systematically with the number of nodes. Wavefunctions exhibit symmetry $\psi_n(-x) = (-)^n \psi(+x)$ if the potential is invariant under reflection: $V(-x) = +V(+x)$.

an eigenfunction increases systematically with energy. In fact, in a one-dimensional potential we can use the number of nodes in an eigenfunction to identify that eigenfunction: $\psi_n(x)$ has n $(n = 0, 1, 2, \ldots)$ nodes and energy E_n. The larger n, the larger E_n. We observe in passing that the wavefunctions look like sine functions with variable argument and amplitude: $\psi(x) \sim A(x) \sin k(x) x$.

The square well and Gaussian potentials are reflection symmetric: $V(-x) = V(+x)$, when the coordinate system is centered at the midpoint of the potential. In these cases the wavefunctions are even or odd under reflection, depending on the number of nodes

$$\psi_n(-x) = (-)^n \psi_n(+x) \qquad n = 0, 1, 2, 3, \ldots . \qquad (29.5)$$

The third potential is not symmetric and its wavefunctions do not have reflection symmetry. We observe that the wavefunction at 7.37 eV varies very slowly over the shallow left-hand part of the potential at $V = 7.0$ eV and more rapidly over the deeper right-hand part of the potential at 0.0 eV. This is a general feature of wavefunctions. Their curvature $(-d^2\psi/dx^2)$ is proportional to the depth of the potential $(= [2m(E - V)/\hbar^2] \, \psi)$. This is just a manifestation of Schrödinger's equation.

We also remark that the wavefunction extends into the classically forbidden region. The range of this penetration is $\kappa^{-1} = 1/\sqrt{2m(V-E)/\hbar^2}$. The smaller $V - E$, the further the wavefunction extends into the classically forbidden region. This is why the wavefunctions become more "delocalized" as the energy increases. This is especially noticeable in the least bound state $\psi_3(x)$ with energy $E_3 = 19.79$ eV in the Gaussian potential. Such loosely bound states are said to be near the "ionization threshold."

We now illustrate how wavefunctions can be used to identify the source of eigenvalues in complicated potentials. In Fig. 29.2 we show two isolated potential wells. The well on the left has unequal asymptotic potentials $V_L = 1.0$ eV, $V_R = 0.0$ eV, a depth of 5.0 eV below the lower asymptotic potential, and a length of 6.0 Å. It supports bound states at -4.367, -2.549, and -0.064 eV. The well on the right has $V_L = 0.0$, $V_R = 1.0$ eV, a depth of 3.0 eV below the lower asymptotic potential, and a length of 4.0 Å. It supports one bound state at -2.028 eV.

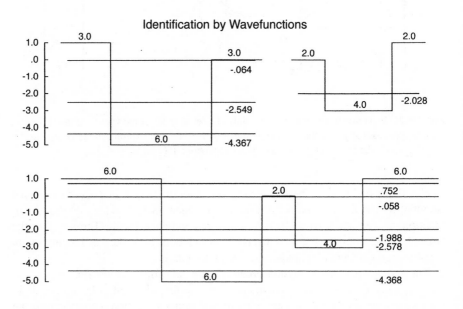

Fig. 29.2   Energy eigenvalues of two isolated wells (top) and a double-well potential formed from them.

When the two wells are combined, as shown in the lower half of Fig. 29.2, there are five bound states, as indicated. Simply by comparing eigenvalues, we expect the states at -4.368, -2.578, and -0.058 eV to be "associated" with the left-hand well and the state with energy -1.988 eV to be associated with the right-hand well. The new state at $+0.752$ eV is a consequence of the boundary conditions ($V_L = V_R = +1.0$ eV).

These identifications can be made more precise by computing the bound state wavefunctions in the double-well potential and comparing these wavefunctions with those in the two isolated wells. This comparison is carried out in Fig. 29.3. All wavefunctions in this figure are normalized to unity and plotted to the same scale. They are also plotted on the horizontal line that defines their energy eigenvalue in the well.

Identification by Wavefunctions

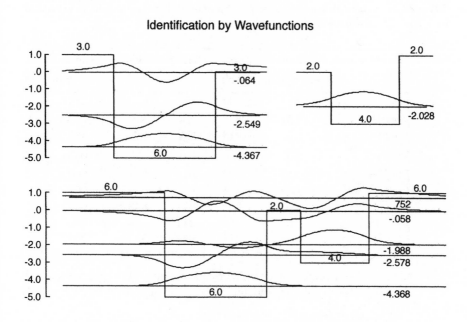

Fig. 29.3   Wavefunctions for the potentials shown in Fig. 29.2

This figure shows that the lowest eigenstate at -4.368 eV is confined entirely to the left-hand well. It differs from the ground state at -4.367 eV in the single-well potential in no discernible way. The double-well state at -2.578 is confined largely, but not entirely, to the left-hand well. In this well it strongly resembles the single-well state at -2.549 eV. The double-well state at -1.988 eV is confined largely to the right-hand well, in which it strongly resembles the ground state wavefunction at $E = -2.028$ in the isolated single-well potential. The double-well state with $E = -0.058$ is not too well confined to the left-hand well. This could have been anticipated simply by looking at the wavefunction in the isolated well at the left. Since its energy (-0.064 eV) is only slightly below the barrier, the amplitude drops off very slowly toward the right. This means that in the double-well potential it will extend well past the barrier whose thickness is only 2.0 Å. Finally, the fifth state in the double-well potential at $E = +0.752$ eV is a consequence of the boundary condition $V_L = V_R = +1.0$ eV. This wavefunction extends over both wells and doesn't resemble any of the wavefunctions in the isolated wells.

We remark that the wavefunctions become increasingly "delocalized" with increasing energy. This means that the probability that the particle is in its "home" potential well decreases and the probability of being elsewhere increases, as the energy increases. These probabilities are summarized in Table 29.1.

Table 29.1 Probability that an electron in either of the isolated single wells, or in the double well, is in various regions of the potential

Wavefunction	Left Asymptotic	Left Well	Barrier	Right Well	Right Asymptotic
$\int |\psi_0^L|^2$	.012	.973	NA	NA	.015
$\int |\psi_1^L|^2$	.051	.876	NA	NA	.073
$\int |\psi_2^L|^2$	.100	.413	NA	NA	.487
$\int |\psi_0^R|^2$	.068	NA	NA	.890	.042
$\int |\psi_0^D|^2$	.011	.972	.016	.001	.000
$\int |\psi_1^D|^2$	.047	.821	.085	.046	.001
$\int |\psi_2^D|^2$	.005	.064	.041	.848	.042
$\int |\psi_3^D|^2$	.109	.450	.211	.174	.056
$\int |\psi_4^D|^2$	.121	.211	.054	.345	.269

Note: Wavefunctions are identified by the potential (L = isolated left well, R = isolated right well, D = double well) and number of nodes (0, 1, 2, 3, 4).

Problem. For an electron in the isolated potential on the left, compute the probability that the electron is *inside* the well ($0 \leq x \leq 6.0$ Å) when it is in: (i) the ground state; (ii) the first excited state; and (iii) the second excited state.

Problem. For an electron in the isolated potential on the right, compute the probability that the electron is the well ($0 \leq x \leq 4.0$ Å) when it is in the ground state.

Problem. For an electron in the double-well potential, compute the probability that it is in the left-hand well and the right-hand well. Do this for each of the five eigenstates. Compare these probabilities with the probabilities computed above for electrons in the isolated wells.

30

Superpositions, Overlaps, and Probabilities

In previous chapters we have encountered an increasingly powerful set of techniques for comparing quantum mechanical systems. In this chapter we present a final powerful method for comparing such systems.

We illustrate these comparison methods by considering the double-well potential made by placing two isolated single-well potentials adjacent to each other (see Fig. 29.2). By "adjacent" we mean that the barrier that separates the two wells is sufficiently thin that there is a nonnegligible tunneling probability between wells. We have asked the question: "How are the states in the double well related to the states in the isolated wells?" We have responded to this question in three ways so far:

1. We have computed the energy eigenvalues in the double well and compared these energies with the eigenvalues in the isolated wells (Fig. 29.2). From this comparison we have concluded that the states ψ_0^D, ψ_1^D, and ψ_3^D are very similar to the states ψ_0^L, ψ_1^L, and ψ_2^L in the left well and state ψ_2^D is similar to ψ_0^R in the right-hand well. The most excited double-well state ψ_4^D is not similar to any eigenstate in either well.

2. We have computed the wavefunctions in the double well and compared them with the wavefunctions in the two isolated wells. This visual comparison (Fig. 29.3) confirms our previous identifications.

3. For each of the five states of the double-well potential we have computed the probability that the electron is in the left- or the right-hand well (Table 29.1). These probabilities were compared with the probabilities of the electron being in the potential well for the two isolated single wells in problems (29.1)–(29.3).

The first two comparison methods above are qualitative in nature ("it looks similar," "it doesn't look the same"). The third verges on being quantitative. We now present a quantitative comparison method.

Usually when we say that one wavefunction is similar to another, or like a few others, we mean that it can be expressed as some linear combination of the others. In this case, we would like to see how similar the wavefunctions ψ_α^D, ($\alpha = 0, 1, 2, 3, 4$) in the double well are to those in the single-well potentials. To do this, we represent ψ_α^D as a linear superposition of the three wavefunctions ψ_i^L, ($i = 0, 1, 2$) in the isolated left-hand potential and the single eigenfunction ψ_0^R in the isolated right-hand well

$$\psi_\alpha^D = \sum_{i=0}^{2} C_{\alpha i}\psi_i^L + D_{\alpha 0}\psi_0^R + \text{"other stuff"}. \tag{30.1}$$

Naturally, we can't expect each (any) of the five double-well wavefunctions to be exactly represented by the four different wavefunctions in the two separated wells. The "other stuff" is what remains after the coefficients $C_{\alpha i}$, $D_{\alpha 0}$ have been chosen to provide a "best fit" to the wavefunction ψ_α^D as a linear superposition of ψ_i^L and ψ_0^R. The "other stuff" is orthogonal to the four wavefunctions ψ_i^L and ψ_0^R.

The standard way to compare wavefunctions is to compute their inner product or overlap integral:

$$\langle \psi_i^L | \psi_\alpha^D \rangle = \int_{-\infty}^{+\infty} \overline{\psi_i^L(x)}\psi_\alpha^D(x)dx . \tag{30.2}$$

Each of these wavefunctions is available as a set of amplitudes $\begin{bmatrix} A \\ B \end{bmatrix}_j$ in successive pieces of the potential, or as a list of values $\psi_i^L(x_k)$, $k = 0, 1, 2, \ldots, P$, where P is a sufficiently large number that this list of $P + 1$ numbers is a good approximation to the smooth function $\psi_i^L(x)$. Such a list of numbers is actually what has been plotted in Fig. 29.3 (one number at each "pixel" along the x-axis). This list of numbers can be treated as a column vector of length $P + 1$, and the overlap integral can be computed by taking the inner product of two $(P + 1) \times 1$ matrices:

$$\int_{-\infty}^{+\infty} \overline{\psi_i^L(x)}\psi_\alpha^D(x)dx \simeq \sum_{k=0}^{P} \overline{\psi_i^L(x_k)}\psi_\alpha^D(x_k)(x_{k+1} - x_k) . \tag{30.3}$$

The larger the number of points, or the finer the mesh on which $\psi_i^L(x_k)$ is evaluated, the closer the right-hand side becomes to the integral on the left. We have computed inner products using the computational method on the right-hand side of (30.3) since the list of numbers representing the wavefunction was already available.

We present overlap integrals in Table 30.1. This includes not only the overlaps between the double-well eigenstates and the single-well eigenstates, but also the single-well eigenstates among themselves.

This table shows clearly that the eigenfunctions in the isolated left-hand well are orthogonal to each other. More precisely, since they have been normalized to unity,

Table 30.1   Overlap integrals of the single-well eigenstates with each other and with double-well eigenstates

	ψ_0^L	ψ_1^L	ψ_2^L	ψ_0^R	ψ_0^D	ψ_1^D	ψ_2^D	ψ_3^D	ψ_4^D
ψ_0^L	1.000	.000	.000	.061	.999	−.025	−.004	−.019	.019
ψ_1^L	.000	1.000	.000	.115	−.020	.983	−.183	.015	.017
ψ_2^L	.000	.000	1.000	.559	.026	.104	.585	−.654	.407
ψ_0^R	.061	.115	.559	1.000	.072	.291	.952	.033	−.023

Note: The double-well eigenstates are mutually orthonormal, so their overlap integrals are not shown.

they are mutually orthonormal:

$$\langle \psi_i^L | \psi_j^L \rangle = \delta_{ij} \quad = \quad 1 \text{ if } i = j \\ = \quad 0 \text{ if } i \neq j \quad . \tag{30.4}$$

The eigenfunctions in the double-well potential are also mutually orthonormal

$$\langle \psi_\alpha^D | \psi_\beta^D \rangle = \delta_{\alpha\beta} . \tag{30.5}$$

If the right-hand well supported more than one eigenfunction, these also would be mutually orthonormal. These results are all consequences of the theorem that eigenfunctions with different eigenvalues *in the same potential* are orthogonal.

We next observe that some overlaps are large while others are small. In fact, wavefunctions that look similar in Fig. 29.3 produce large overlaps

$$\begin{aligned} \langle \psi_0^D | \psi_0^L \rangle &= .999 , \\ \langle \psi_1^D | \psi_1^L \rangle &= .983 , \\ \langle \psi_2^D | \psi_0^R \rangle &= .952 , \\ \langle \psi_3^D | \psi_2^L \rangle &= -.654 . \end{aligned} \tag{30.6}$$

These overlaps decrease with increasing energy. This comes about physically because as the energy increases the electron becomes more delocalized and has a higher probability of tunneling from one well to another. As a result, the double-well eigenstates become less like single-well eigenstates with increasing energy.

Problem. Why is the overlap $\langle \psi_3^D | \psi_2^L \rangle$ negative while the other large overlaps are positive?

If the basis states ψ_i^L, ψ_0^R from which we attempt to construct double-well eigenstates ψ_α^D were all orthonormal, the overlap integrals would have a natural interpretation as probability amplitudes, and their absolute squares as probabilities. That is, $\langle \psi_2^D | \psi_2^L \rangle = 0.585$ would be the probability amplitude for finding the electron in the left well eigenstate ψ_2^L when it is in the double-well eigenstate ψ_2^D, and $|\langle \psi_2^D | \psi_2^L \rangle|^2 = 0.342$ would be the probability. However, the single-well eigenstates are not orthogonal (e.g., $\langle \psi_2^L | \psi_0^R \rangle = 0.559$), so we must be very careful about this interpretation. To emphasize this point, we observe that the probability for finding an electron in the state ψ_0^R if it is in the double-well state ψ_2^D, $\Pr(\psi_0^R|\psi_2^D)$, is $|\langle\psi_0^R|\psi_2^D\rangle|^2 = (0.952)^2 = 0.906$. Combining this with the previous result, we find

$$\Pr(\psi_2^L|\psi_2^D) + \Pr(\psi_0^R|\psi_2^D) = 0.342 + 0.906 = 1.248 > 1.0 \, . \tag{30.7}$$

The sum of these probabilities exceeds 1.0!

It is clear that we must understand the probabilistic interpretation at a deeper level when the states in terms of which we wish to expand a set of wavefunctions are neither orthogonal nor complete. To do this, we let $\psi_\alpha = |\alpha\rangle$ (e.g., ψ_α^D) represent a set of states. We wish to expand these states in terms of another set of states, $\phi_\mu = |\mu\rangle$ (e.g., ψ_i^L, ψ_0^R), which are not necessarily orthogonal or complete. The expansion will have the form (see (30.1))

$$|\alpha\rangle = \sum_\mu |\mu\rangle C_{\mu\alpha} + \text{``other stuff''} \, . \tag{30.8}$$

If the states $|\mu\rangle$ are not complete, the state $|\alpha\rangle$ cannot be completely expressed in terms of the states $|\mu\rangle$. The function labeled "other stuff" is a residual function, which is left over, after we have chosen the coefficients $C_{\mu\alpha}$ so that the states $|\mu\rangle$ provide the "best possible approximation" to $|\alpha\rangle$.

Best in what sense? The most useful interpretation is "best in the least squares sense." That is, we wish to minimize the length of the residual wavefunction. This is done by minimizing the square of the difference by appropriate choice of the expansion coefficients

$$\text{Minimize} \left(|\alpha\rangle - \sum_\mu |\mu\rangle C_{\mu\alpha} \right)^2 . \tag{30.9}$$

This is done, for each state $|\alpha\rangle$, by taking derivatives with respect to $C_{\mu\alpha}$ and setting the result equal to zero. This leads immediately to the following simple equation for the coefficients $C_{\nu\alpha}$:

$$\sum_\nu \langle\mu|\nu\rangle C_{\nu\alpha} = \langle\mu|\alpha\rangle \, . \tag{30.10}$$

Illustration. For the state $\psi_\alpha = \psi_2^D$, equation (30.10) is

$$\begin{bmatrix} 1.000 & 0.000 & 0.000 & 0.062 \\ 0.000 & 1.000 & 0.000 & 0.115 \\ 0.000 & 0.000 & 1.000 & 0.559 \\ 0.062 & 0.115 & 0.559 & 1.000 \end{bmatrix} \begin{bmatrix} C_{L,0}^{D,2} \\ C_{L,1}^{D,2} \\ C_{L,2}^{D,2} \\ C_{R,0}^{D,2} \end{bmatrix} = \begin{bmatrix} -0.004 \\ -0.183 \\ 0.585 \\ 0.952 \end{bmatrix} . \tag{30.11}$$

All overlaps in (30.11) can be read from Table 30.1.

The coefficients $C_{\nu\alpha}$ can be determined by computing the inverse of the overlap matrix $\langle\mu|\nu\rangle$, which we denote $\langle\nu|\mu\rangle = \langle\mu|\nu\rangle^{-1}$ and multiplying

$$C_{\nu\alpha} = \langle\mu|\nu\rangle^{-1}\langle\mu|\alpha\rangle = \langle\nu|\mu\rangle\langle\mu|\alpha\rangle \, . \tag{30.12}$$

Illustration. For ψ_2^D again

$$
\begin{bmatrix} C_{L,0}^{D,2} \\ C_{L,1}^{D,2} \\ C_{L,2}^{D,2} \\ C_{R,0}^{D,2} \end{bmatrix} = \begin{bmatrix} 1.006 & 0.011 & 0.051 & -0.092 \\ 0.011 & 1.020 & 0.096 & -0.171 \\ 0.051 & 0.096 & 1.467 & -0.835 \\ -0.092 & -0.171 & -0.835 & 1.492 \end{bmatrix} \begin{bmatrix} -0.004 \\ -0.183 \\ 0.585 \\ 0.952 \end{bmatrix} = \begin{bmatrix} -0.063 \\ -0.293 \\ 0.046 \\ 0.963 \end{bmatrix}
$$

With these coefficients, we can compute the length of the "other stuff":

$$
\left(|\alpha\rangle - \sum_{\mu,\nu} |\nu\rangle \langle\nu|\mu\rangle \langle\mu|\alpha\rangle \right)^2 = \langle\alpha|\alpha\rangle - \sum_{\mu\nu} \langle\alpha|\mu\rangle \langle\mu|\nu\rangle \langle\nu|\alpha\rangle . \tag{30.13}
$$

The right-hand side is the probability that an electron in the state $|\alpha\rangle$ is not in the subspace spanned by the nonorthogonal states $|\mu\rangle$. Therefore, the probability that the electron *is* in the space spanned by the states $|\mu\rangle$ is

$$
\Pr = \sum_{\mu\nu} \langle\alpha|\mu\rangle \langle\mu|\nu\rangle \langle\nu|\alpha\rangle = \sum_{\mu} \langle\alpha|\mu\rangle C_{\mu\alpha} . \tag{30.14}
$$

Illustration. For ψ_3^D this is the inner product of the overlaps $\langle\psi_3^D|\psi_i^L\rangle$, $\langle\psi_3^D|\psi_0^R\rangle$ with the coefficients $C_{L,i}^{D,3}$, $C_{R,0}^{D,3}$ is

$$
\Pr = (-0.019, 0.015, -0.654, 0.033) \begin{pmatrix} -0.055 \\ -0.093 \\ -0.986 \\ 0.593 \end{pmatrix} = 0.664 . \tag{30.15}
$$

This result tells us that the state ψ_3^D is not very well described as a linear superposition of the four single-well states.

It is still useful to estimate the probability that if an electron is in state $|\alpha\rangle$, it is in state $|\mu\rangle$ as the square of the overlap integral:

$$
\Pr(|\mu\rangle||\alpha\rangle) = |\langle\mu|\alpha\rangle|^2 \le 1.0 . \tag{30.16}
$$

For normalized states, this is always bounded above by $+1.0$, by the Schwartz inequality. However, the sum of these probabilities may exceed $+1.0$ since the wavefunctions $|\mu\rangle, |\nu\rangle$ are not orthogonal. What must be true is that the weighted sum of the probabilities $(\Pr(|\mu\rangle||\alpha\rangle) \times \langle\mu|\mu\rangle^{-1})$ and the interference terms $\langle\alpha|\mu\rangle \langle\mu|\nu\rangle^{-1} \langle\nu|\alpha\rangle$ ($\mu \ne \nu$) must sum to the probability that an electron in the state $|\alpha\rangle$ lies in the subspace spanned by the states $|\nu\rangle$:

$$
\sum_{\mu} \Pr(|\mu\rangle||\alpha\rangle)) \times \langle\mu|\mu\rangle^{-1} + \sum_{\mu\ne\nu} \sum_{\nu} \langle\alpha|\mu\rangle \langle\mu|\nu\rangle \langle\nu|\alpha\rangle
$$
$$
= \sum_{\mu} \sum_{\nu} \langle\alpha|\mu\rangle \langle\mu|\nu\rangle \langle\nu|\alpha\rangle \le 1.0 . \tag{30.17}
$$

In Table 30.2 we present the coefficients $C_{\mu\alpha}$ for the expansion of the five eigenstates of the double-well potential in terms of the four isolated single-well states. In Table 30.3 we present the squares of the overlaps $|\langle \psi_\alpha^D | \psi_i^L \rangle|^2$, $|\langle \psi_\alpha^D | \psi_0^R \rangle|^2$, which can be interpreted as probabilities. We also present the probabilities that an electron in this double-well eigenstate lies in the subspace spanned by the single-well states. The results show that the lowest three double-well states are essentially completely described by the single-well states (probabilities 1.000, 1.000, and 0.998), ψ_3^D is poorly described (Pr $= 0.664$), and the most energetic state ψ_4^D is very poorly described (Pr $= 0.263$). Although these numbers present no surprises, they make quantitative what we understood previously at a qualitative (or intuitive) level.

Table 30.2 Expansion coefficients $C_{\mu\alpha}$ of the five eigenstates in the double-well potential in terms of the four eigenstates of the isolated single-well potentials

	ψ_0^L	ψ_1^L	ψ_2^L	ψ_0^R
ψ_0^D	.999	−.020	.027	−.002
ψ_1^D	−.037	.962	.002	.182
ψ_2^D	−.063	−.293	.046	.963
ψ_3^D	−.055	−.053	−.986	.593
ψ_4^D	.043	.060	.619	−.379

Table 30.3 Probability of finding a single-well eigenstate in each of the five double-well eigenstates

	ψ_0^L	ψ_1^L	ψ_2^L	ψ_0^R	Probability in Subspace
ψ_0^D	.998	.000	.001	.005	1.000
ψ_1^D	.001	.967	.011	.085	1.000
ψ_2^D	.000	.033	.342	.906	0.998
ψ_3^D	.000	.000	.427	.001	0.664
ψ_4^D	.000	.000	.166	.001	0.263

31

Symmetry and Wavefunctions

The results presented in Fig. 29.3 and Table 30.3 show that eigenfunctions in multiple potential wells show clear signs of their parentage when the wells are different and the energies are widely separated.

To determine what happens when eigenvalues become degenerate, in particular in the case of a potential consisting of identical equally spaced wells, we compute eigenstates for potentials with two-, three-, and four-identical wells. The eigenfunctions for the two well potential are shown in Fig. 31.1. The single potential well of width 7.0 Å and depth 5.0 eV supports three bound states. The double-well potential formed from two such identical wells separated by a barrier (0.0 eV) of width 2.0 Å supports six bound states. Plotting all six wavefunctions on this double-well potential produces a hard-to-decipher mess. Therefore we have duplicated the double-well potential and plotted half the eigenfunctions on one copy, the other half on the duplicate copy.

On the left we plot the eigenfunctions associated with the lower member of each doublet. These eigenfunctions are all symmetric under reflection about the midpoint of the symmetric potential: $\Phi(-x) = +\Phi(+x)$. On the right we plot eigenfunctions associated with the more energetic of the two members of each doublet. These functions are all odd under reflection: $\Phi(-x) = -\Phi(+x)$. In both cases, displacements x are measured from the midpoint of the potential, so that the potential is symmetric under reflection: $V(-x) = +V(+x)$.

At this point it is worthwhile to make a number of observations about these wavefunctions and their properties. These observations are valid, with suitable modifications, for the eigenfunctions in potentials with N identical equally spaced wells. It

Symmetry and Wavefunctions

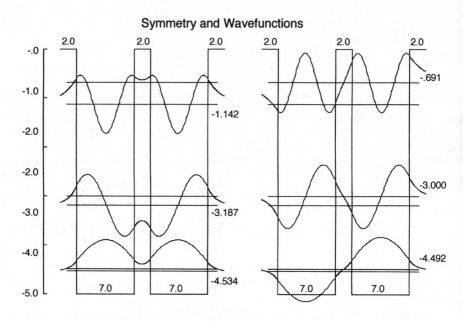

Fig. 31.1   Wavefunctions for the double-well potential. *Left*, Even parity states. *Right*, Odd parity states. Even parity state is lowest energy member of each doublet. The symmetry type of the even parity state alternates between symmetric (s: $k = 0, 2, 4, \ldots$) and antisymmetric (a: $k = 1, 3, 5, \ldots$).

is necessary here to distinguish between symmetry (an algebraic property) and parity (a geometric property):

1. The number of nodes (zero crossings) of these wavefunctions increases systematically with energy: $E_{n+1} > E_n$. The "ground state" has zero nodes and the nth excited state has n nodes.

2. All states in this potential are even or odd. The parity alternates with energy according to

$$\psi_n(-x) = (-)^n \psi_n(+x) \,, \tag{31.1}$$

where the state $\psi_n(x)$ has n nodes ($n = 0, 1, 2, \ldots$).

3. Each state in the double-well potential is a linear combination of states in the single-well potential. If $\phi_{l,k}$ represents an eigenstate with k nodes in the isolated single well on the left and $\phi_{r,k}$ is the corresponding state for the isolated well on the right, then the two states $\psi_{s,k}, \psi_{a,k}$ in the kth doublet of the double-well

potential are

$$\begin{bmatrix} \psi_{s,k} \\ \psi_{a,k} \end{bmatrix} = \frac{1}{\sqrt{2}} \begin{bmatrix} 1 & 1 \\ -1 & 1 \end{bmatrix} \begin{bmatrix} \phi_{l,k} \\ \phi_{r,k} \end{bmatrix}. \tag{31.2}$$

The state $\psi_{s,k} = (\phi_{l,k}+\phi_{r,k})/\sqrt{2}$ is a symmetric linear combination of the state $\phi_{l,k}$ with k nodes in the isolated left well and $\phi_{r,k}$ with k nodes in the isolated right well. The state $\psi_{a,k} = (-\phi_{l,k} + \phi_{r,k})/\sqrt{2}$ is an antisymmetric linear combination of the two single-well states. For deep wells the symmetric and antisymmetric states are approximately eigenstates of the two-well potential.

4. These linear combinations can be determined by inspection of the wavefunctions in Fig. 31.1. For example, in the ground state doublet ($k = 0$) the symmetric state $(1, 1)/\sqrt{2}$ is the sum of a wavefunction with no nodes in the left well and its counterpart in the right-hand well. The antisymmetric state $(-1, 1)/\sqrt{2}$ is the difference of these two wavefunctions. The same is true for all other doublets.

5. In the ground state doublet, the symmetric state has even parity and the antisymmetric state has odd parity. In the $k = 1$ doublet, the symmetric state has odd parity and the antisymmetric state has even parity. The $k = 2$ ($4, 6, \dots$) doublet is like the $k = 0$ doublet and all odd doublets behave similarly:

$$\begin{aligned} \psi_{s,k}(-x) &= (-)^k \psi_{s,k}(+x), \\ \psi_{a,k}(-x) &= (-)^{k+1} \psi_{a,k}(+x). \end{aligned} \tag{31.3}$$

6. In the lowest doublet ($k = 0$) the symmetric state has lower energy than the antisymmetric state. In the next doublet ($k = 1$) the order is reversed. The order alternates with k.

7. The ordering of doublet states (by energy) can be done by counting nodes. The single-well wavefunctions $\phi_{l,k}, \phi_{r,k}$ each have k nodes, and asymptotically approach zero with the same (k even) or opposite (k odd) sign, depending on k. The symmetry type (s, a) does not force, or does force, an extra node depending on k. For example, in the $k = 1$ doublet $\phi_{l,k}, \phi_{r,k}$ both have one node. The symmetric linear combination forces a node in the barrier between them, while the antisymmetric combination does not force this extra node. This is why $(\phi_{l,1} - \phi_{r,1})/\sqrt{2}$ has two nodes, has even parity, and has lower energy than $(\phi_{l,1} + \phi_{r,1})/\sqrt{2}$, which has three nodes, odd parity, and higher energy.

The probability distributions for the symmetric and antisymmetric states in the same doublet of the symmetric double-well potential are difficult to distinguish. The squares of the wavefunctions shown in Fig. 31.1 are shown in Fig. 31.2. It is clear that the wavefunctions can be easily distinguished, but their probability distributions cannot. It is for this reason that we were unable to distinguish the two peaks in a transmission probability doublet through their probability distributions alone (Part II, chapter 16).

Symmetry and Probability Distributions

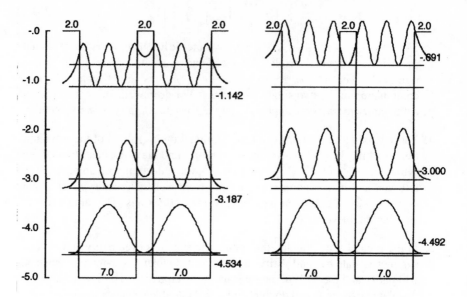

Fig. 31.2 Probability distributions for the wavefunctions shown in Fig. 12.1. Probability distribution functions for the two states in a multiplet are much harder to distinguish than the corresponding wavefunctions.

In the case of potential wells with three or more identical equally spaced wells, the eigenfunctions are symmetrized linear combinations of the eigenfunctions of a single well, but the method of symmetrization is not quite as simple as in the two well case. We show eigenfunctions for a three well potential in Fig. 31.3 and for a four well potential in Fig. 31.4.

In the three well potential, eigenfunctions occur in triplets. Within each triplet we plot the three eigenfunctions in terms of increasing energy, from left to right. The triplets are plotted in order of energy: the lowest energy triplet on the lowest line; the next triplet on the line above, and so on. This means the energy of the eigenfunctions increases from left to right, bottom to top, in this figure. All eigenfunctions have been plotted over a representation of the three well potential, so the peaks and nodes can be identified with the various parts of the potential.

We can make the following observations from this figure.

1. The number of nodes increases systematically with energy. In the lowest energy triplet (lowest line) the eigenfunctions have zero, one, and two nodes. In the next triplet they have three, four, and five nodes. In the next they have six, seven, and eight nodes.

Symmetry and Wavefunctions

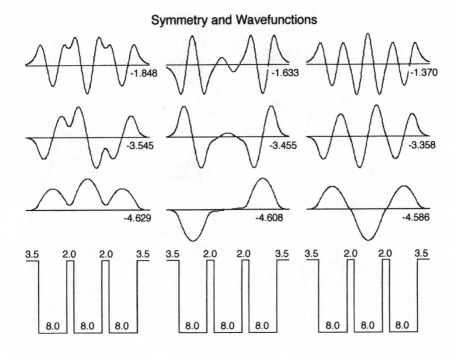

Fig. 31.3 Eigenstates for three-well potential. Potential is shown in bottom row. Successive rows from bottom are for triplets of increasing energy. Energy within each multiplet increases from left to right.

2. The eigenfunctions have even or odd parity according to

$$\psi_n(-x) = (-)^n \psi_n(+x) , \qquad (31.4)$$

where n is the number of nodes in the wavefunction and the origin is at the center of the potential, which obeys $V(-x) = +V(+x)$.

3. In the lowest triplet the eigenfunctions in order of increasing energy have parity e, o, e (*even, odd*). In the next triplet they are o, e, o. This pattern repeats itself. This means that the lowest member of a triplet does not always have even parity.

4. In the ground state triplet, the lowest eigenfunction is a symmetric linear combination of three eigenfunctions for the three isolated wells

$$\psi_{s,0} = \frac{1}{\sqrt{3}} (\phi_{l,0} + \phi_{m,0} + \phi_{r,0}) , \qquad (31.5)$$

where $\phi_{l,k}$ is an eigenfunction with k nodes in the left well, and so on. The factor $1/\sqrt{3}$ is for normalization purposes. The next state in the ground state multiplet

Symmetry and Wavefunctions

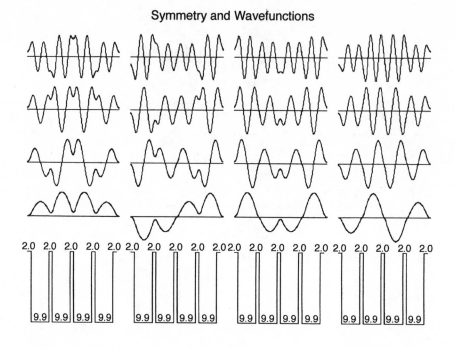

Fig. 31.4 Eigenstates for four-well potential. Organization of eigenfunctions follows pattern described in Fig. 31.3

is a different linear combination of these three single-well eigenfunctions:

$$\psi_{a,0} = \frac{1}{\sqrt{2}} \left(-\phi_{l,0} + \phi_{r,0} \right) . \tag{31.6}$$

The third eigenfunction is yet another linear combination:

$$\psi_{i,0} = \frac{1}{\sqrt{6}} \left(\phi_{l,0} - 2\phi_{m,0} + \phi_{r,0} \right) . \tag{31.7}$$

Here i stands for intermediate symmetry. These relations can be inferred by inspection of the wavefunctions for the ground state multiplet. However, they hold true for all multiplets. For example, for the kth multiplet we find

$$\begin{bmatrix} \psi_{s,k} \\ \psi_{a,k} \\ \psi_{i,k} \end{bmatrix} = \begin{bmatrix} \frac{1}{\sqrt{3}} & \frac{1}{\sqrt{3}} & \frac{1}{\sqrt{3}} \\ -\frac{1}{\sqrt{2}} & 0 & \frac{1}{\sqrt{2}} \\ \frac{1}{\sqrt{6}} & -\frac{2}{\sqrt{6}} & \frac{1}{\sqrt{6}} \end{bmatrix} \begin{bmatrix} \phi_{l,k} \\ \phi_{m,k} \\ \phi_{r,k} \end{bmatrix} . \tag{31.8}$$

As in the two well case, the energy of these symmetrized eigenfunctions alternates with bands. For example, the symmetric linear combination $(1,1,1)/\sqrt{3}$

has lowest energy in the multiplets with $k = 0, 2, 4, \ldots$ and highest energy in the odd bands $k = 1, 3, \ldots$.

Fig. 31.4 tells a similar story for a potential consisting of four identical wells. To create this figure we have deepened and widened the potential so that more bound states exist. The presentation of wavefunctions in Fig. 31.4 follows the same format as used in Fig. 31.3. Each quartet is shown on a separate line. Energy increases from left to right, bottom to top. So also does the number of nodes, from 0 to 15. The wavefunctions are also parity eigenstates: $\phi_n(-x) = (-)^n \phi_n(+x)$. Each energy eigenfunction $\psi_{\alpha,k}$, ($\alpha = 1, 2, 3, 4$) in the kth multiplet is a linear combination of single-well eigenfunctions $\phi_{\beta,k}$ in the various wells $\beta = 1, 2, 3, 4$. The index α on $\psi_{\alpha,k}$ describes symmetry type including symmetric as well as three additional types. These linear combinations can be inferred from Fig. 31.4. In particular, the coefficients are easiest to deduce from the ground state multiplets. The result is

$$
\begin{bmatrix} \psi_{1,k} \\ \psi_{2,k} \\ \psi_{3,k} \\ \psi_{4,k} \end{bmatrix} = \frac{1}{2} \begin{bmatrix} 1 & 1 & 1 & 1 \\ -1 & -1 & 1 & 1 \\ 1 & -1 & -1 & 1 \\ -1 & 1 & -1 & 1 \end{bmatrix} \begin{bmatrix} \phi_{1,k} \\ \phi_{2,k} \\ \phi_{3,k} \\ \phi_{4,k} \end{bmatrix} . \tag{31.9}
$$

As in the two and three well cases, the order of levels alternates between multiplets. That is, the symmetric linear combination has even parity and is energetically the lowest of the four eigenfunctions in multiplets with $k = 0, 2, 4, \ldots$ and has odd parity and is energetically the highest of the four eigenfunctions in the odd quartets $k = 1, 3, 5, \ldots$. These observations can all be made by counting nodes and asymptotes in the single-well eigenfunctions and deciding how many additional nodes are forced by the different symmetries $\alpha = 1, 2, 3, 4$ identified by the matrices (31.8) for three wells and (31.9) for four.

32

Transmission Resonances and Bound States

We have used intuitive arguments based on "resonance conditions" twice so far to determine important properties of potentials. We have used a resonance argument in Part II, chapter 16, equation (16.4) to estimate the location of peaks in the transmission probability function, $T(E)$, for an electron incident on a double barrier in the classically forbidden regime. We have used essentially the same argument in Part III, chapter 23, equation (23.3) to estimate the energy at which bound states occur.

If the same argument can be used to locate bound states and peaks in the transmission spectrum, surely they must be closely related.

To verify that this is the case, in Fig. 32.1 we locate the peaks of $T(E)$ for an electron incident on a series of double-barrier potentials with barriers of increasing thickness. We also locate the energy eigenvalues for the potential well associated with this double barrier. We see that the peaks and bound states occur at almost the same energies in these potentials subject to two different boundary conditions. As the barrier thickness increases from 1.0 Å to 2.0 Å to 3.0 Å, the lowest resonance increases from 0.345 to 0.373 to 0.376 eV, approaching the lowest bound state energy at 0.377 eV in the corresponding single-well potential. The energy at which the second resonance occurs also approaches the energy of the first excited bound state ψ_1 from below as the barriers get thicker. However, the third resonance approaches the energy E_2 of the third bound state (second excited state ψ_2), 3.130 eV, from above.

The relation between bound states and resonant scattering states can be made even more striking by comparing their associated probability distribution functions. This comparison is presented in Fig. 32.2. In this figure we show a sequence of three double-barrier potentials. The barriers have the same height (5.0 eV) and constant separation (8.0 Å). The only difference between the three double-barrier potentials is

Scattering Resonances & Bound States

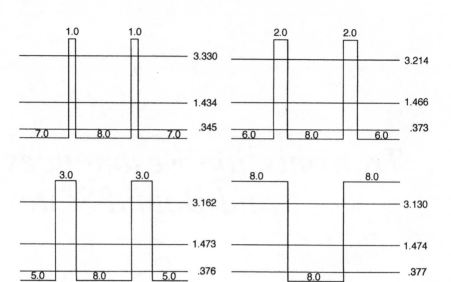

Fig. 32.1 Energies of transmission peak resonances for a double barrier and bound state energy eigenvalues for the corresponding single-well potential.

the thickness of the barriers: they are 2.0 Å, 4.0 Å, and 6.0 Å thick. Increasing the thickness serves to sharpen the transmission resonance peaks.

For each double barrier we plot the associated probability density for each of the transmission resonances following procedures developed in Part II, chapter 16. Since these are scattering states, the probability density is unnormalizable. The densities are all plotted with the same normalization: the integrated probability density is +1 for the region shown, which has a total length of 24.0 Å.

In addition to the three double barriers, we show the corresponding potential well of depth 5.0 eV and width 8.0 Å. The normalized probability density for the three bound states is shown over the horizontal line indicating that bound state.

For the unbound states, the probability density in the asymptotic left- and right-hand regions is constant and equal. This comes about because the wavefunctions in the left- and right-hand regions are

$$
\begin{aligned}
\Phi_L(x) &= e^{+ikx} + Re^{-ikx}, \\
\Phi_R(x) &= Te^{+ikx}.
\end{aligned}
\tag{32.1}
$$

At the peak in the transmission probability, $|T| = 1$, $R = 0$, and $|\Phi_L(x)|^2 = |\Phi_R(x)|^2 = 1$.

We next observe that, at the peak, the ratio of the maximum of the probability density inside the well to the constant density outside the barriers is very large. In

Scattering Resonances & Bound States

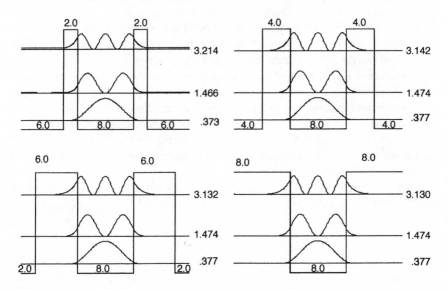

Fig. 32.2 Probability density for a series of three double-barrier potentials with increasingly thick barriers ($D = 2.0, 4.0, 6.0$ Å) and probability distributions for eigenstates in the corresponding single-well potential.

fact, it increases with the width of the barrier and also the depth of the state within the well. To put this into a different perspective, the ratio of the average density outside the barrier to the average probability density between the barriers decreases rapidly with D, the barrier thickness, and $\sqrt{V - E}$, the depth of the transmission resonance beneath the top of the barrier. We have seen this kind of behavior before, and in view of these experiences, we might guess that the logarithm of this ratio is related to the action

$$\log \left[\frac{\text{Average probability density outside barrier}}{\text{Average probability density in well}} \right] = - \int_{c.f.} \sqrt{\frac{2m(V(x) - E)}{\hbar^2}} \, dx$$

(32.2)

where the integral extends over the classically forbidden region. We will not explore this relation at present.

Fig. 32.2 shows clearly that as the barrier thickness increases, the resonance energies in the double barrier approach the bound state energies of the associated single-well potential. The probability densities of the scattering states inside the well converge to the probability distributions of the bound states within the well. Even the exponentially decaying probability density inside the barriers approaches the probability distribution of the bound states in the classically forbidden left- and right-hand regions. It is clear from Fig. 32.2 that there is a very close relation between scattering states at transmission resonances of a double (or multiple) barrier and bound states in the corresponding potential well.

In the previous chapter we stressed that the identification of states should be done through comparison of wavefunctions. We encounter a small problem in attempting such a comparison here. One-dimensional bound state wavefunctions can always be made real. However, the scattering states can never be made real. For this reason the best we can do is to compare the real and imaginary parts of the scattering state wavefunctions with their bound state counterparts. Such a comparison is presented in Fig. 32.3.

This comparison leaves no doubt at all about the relation between scattering state resonances and their bound state counterparts. As the barriers get thicker, the width of the resonance peak shrinks, the ratio of wavefunction amplitude in the well to that outside the barrier increases, and the ratio of imaginary part of the wavefunction to the real part shrinks to zero.

Problem. Test the conjecture that the ratio of the probability density outside the barriers to the average probability density inside the well behaves like (32.2).

Scattering Resonances & Bound States

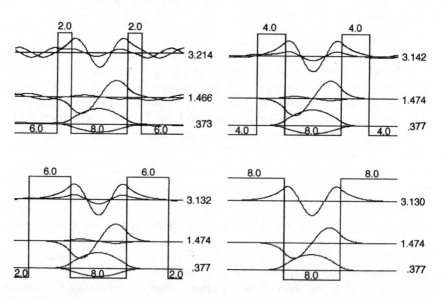

Fig. 32.3 Real and imaginary parts of scattering state wavefunctions at transmission resonances for a sequence of double-barrier potentials with increasingly thick barriers ($D = 2.0$, 4.0, 6.0 Å). These should be compared with the real wavefunctions for the eigenstates of the corresponding single-well potential.

33

Creation of Bound States

In chapter 25 we saw that new bound states "fall into" the top of a potential as the potential is deepened or widened (see Figs. 25.1–25.5). Conversely, they are pushed out of the mouth of the potential as these parameters are made smaller. These phenomena occur for all potentials. New bound states are created and destroyed at the top of a potential well as the appropriate parameter, $A/h = \oint pdq/2\pi\hbar$, is increased or decreased.

In this chapter we will try to understand exactly how this process takes place by studying how the wavefunctions change during the creation of a new bound state. We do this by computing the wavefunction for a square well potential for a variety of potential heights. In particular, we will follow one of the states defined by a dotted curve (resonance peak, $E > V$) and its solid continuation (bound state, $E < V$) in Fig. 25.1. That is, we compute a wavefunction at one of the resonance peaks ($E > V$), then raise the value of the potential and watch how the unbound state is transformed into a bound state.

We will take snapshots of the wavefunction at four stages during this process. The nth resonance peak becomes a bound state when the well depth is $V_n = \frac{\hbar^2}{2m}\left(\frac{n\pi}{L}\right)^2$. We will compute the wavefunctions for a potential well of width 8.0 Å. We set $V_L = V_R = 0.0$ eV. The third resonance peak becomes a bound state when the depth of the potential is $-V_n = -5.284$ eV. For deeper wells this state is bound; for shallower wells it is a resonance.

We choose four well depths: $V = -7.0, -5.4, -5.25$, and -4.0 eV. For $V = -7.0 \ll -V_3$ this bound state exists at -0.875 eV. It is the lowest curve in Fig. 33.1. For $V = -5.4$ eV, just slightly deeper than $-V_3$, this state is very weakly bound at

Scattering Resonances & Bound States

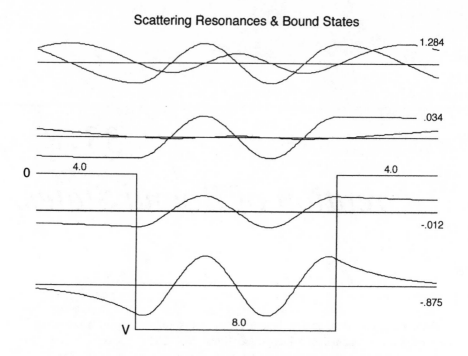

Fig. 33.1 Wavefunctions for a square well potential. *Top to bottom*, $V = -4.0 \gg -V_3$ and $V = -5.25 = -V_3 + \epsilon$ showing real and imaginary parts. Then $V = -5.4 = -V_3 - \epsilon$ and $V = -7.0 \ll -V_3$, showing wavefunction being compressed into the classically allowed region. For this potential $L = 8.0$ Å, $n = 3$ and $V_3 = \frac{\hbar^2}{2m} \left(\frac{n\pi}{L} \right)^2 = 5.284$ eV.

-0.012 eV. Although the wavefunction decays in the classically forbidden region, the decay range is very long.

At $V = -5.25$ eV, slightly above the transition, this state is a resonance at $+0.034$ eV. In the well the imaginary part of the wavefunction is very small compared with the real part. Outside the well the real and imaginary parts of the wavefunction have equal amplitudes. For $V = -4.0$ eV the resonance occurs at $+1.284$ eV. In this state, the ratio of the imaginary to real part of the wavefunction inside the well is substantially larger than for this state at $+0.034$ eV.

Reversing this sequence, we see that as the well depth increases from -4.0 eV to -7.0 eV, the ratio of the imaginary part of the wavefunction to the real part inside the well shrinks to zero as $V \rightarrow -V_3 = -5.284$ eV. As this occurs, the wavelength of the wavefunction outside the barrier approaches ∞. The wavefunction outside the barrier becomes a constant at $-V_3$. Below $-V_3$ the wavefunction begins to decay exponentially in the (now) classically forbidden region. In addition, the wavefunction is now real everywhere. The deeper the well becomes, the more the wavefunction is squeezed out of the classically forbidden region.

At the transition the probability density $|\psi(x)|^2$ has a particularly striking form. It has constant value everywhere outside the well. Inside the well it is the square of a cosine:

$$
\begin{array}{ccc}
x \leq 0 & 0 \leq x \leq L & L \leq x \\
|\psi(x)|^2 = 1 & |\psi(x)|^2 = \cos^2\left(\frac{n\pi x}{L}\right) & |\psi(x)|^2 = 1 \;.
\end{array}
\tag{33.1}
$$

The probability density at the transition is "a ripple in the continuum."

This qualitative understanding can be made more precise by a simple calculation. We consider first the scattering states for the square well potential. We hold the bottom of the potential at $V = 0$ and choose $V_L = V_R = V$. On the resonance, the condition $kL = n\pi$ is satisfied. We choose as wavefunction in this region

$$
\begin{array}{ccc}
x \leq 0 & 0 \leq x \leq L & L \leq x \\
e^{ik'x} & A\cos kx + B\,e^{ikx} & C\,e^{ik'(x-L)} \;,
\end{array}
\tag{33.2}
$$

where as usual $k = \sqrt{2mE/\hbar^2} = n\pi/L$ and $k' = \sqrt{2m(E-V)/\hbar^2}$. The two continuity conditions are easily imposed:

$$
\begin{array}{ccc}
& \text{at } x = 0 & \text{at } x = L \\
\Phi \text{ continuous} & 1 = A + B & (-)^n A + (-)^n B = C \\
\Phi' \text{ continuous} & ik' = ikB & (-)^n ikB = ik'C \;.
\end{array}
\tag{33.3}
$$

It is then immediate that the wavefunction in the asymptotic right-hand region is $\Phi_R(x) = (-)^n e^{ik'(x-L)}$ and the wavefunction in the well is

$$
\begin{aligned}
\Phi_{\text{well}}(x) &= \left(1 - \frac{k'}{k}\right)\cos\,kx + \frac{k'}{k}e^{ikx} \\
&= \cos\,kx + i\frac{k'}{k}\sin\,kx \;.
\end{aligned}
\tag{33.4}
$$

This shows that the ratio of imaginary to real part of the wavefunction approaches zero as the transition is reached:

$$
\frac{Im\ \Phi(x)}{Re\ \Phi(x)} = \sqrt{1 - (V/E_n)} \xrightarrow{V\uparrow E_n} 0 \;.
\tag{33.5}
$$

At the transition the wavefunction is constant in the asymptotic left- and right-hand regions, and a cosine in the well

$$
\begin{array}{ccc}
x \leq 0 & 0 \leq x \leq L & L \leq x \\
1 & \cos\frac{n\pi x}{L} & (-)^n 1 \;.
\end{array}
\tag{33.6}
$$

On the bound state side of the transition the wavefunctions are

$$
\begin{array}{ccc}
x \leq 0 & 0 \leq x \leq L & L \leq x \\
Ae^{\kappa x} & \cos(kx - \phi) & (-)^n Ae^{-\kappa(x-L)} \;.
\end{array}
\tag{33.7}
$$

The conditions that determine A, k, and ϕ can be read from the continuity conditions at $x = 0$, $x = L$. The basic results are as follows: As V increases, ϕ increases from 0 to $\pi/2$ and k increases from $n\pi/L$ to $(n+1)\pi/L$. For large values of V the wavefunction is squeezed out of the classically forbidden regions ($A \to 0, \kappa \to 0$) and the wavefunction in the well becomes

$$\cos\left(\frac{(n+1)\pi x}{L} - \frac{\pi}{2}\right) = \sin\frac{(n+1)\pi x}{L} . \tag{33.8}$$

Although these calculations have been done for the square well potential, the results are qualitatively the same for all one-dimensional potentials. These phenomena can be summarized in the following colorful way. As we sit on a transmission peak and watch the potential get wider and/or deeper ($\oint pdq$ increases) we slide down toward the transition on a resonance that becomes increasingly narrow. At the transition, the mouth of the well reaches out and seizes the resonance from the continuum, and digests it as a bound state. As the bound state sinks to the bottom of the well ($E \ll V_{L,R}$), it gets squeezed out of the classically forbidden region and into the interior of the well.

34

Quantum Engineering

For a variety of reasons it is useful to fabricate devices that consist of many identical copies of some potential. The energy band structure of an N-identical well device consists of N-tuplets of levels spread out around the single well levels.

For technical reasons it is often desirable to dope the device with an "impurity" to create isolated levels in N-well devices. How does one choose the impurity to provide the desired characteristics?

To make this question more explicit, we consider a single well labeled A in Fig. 34.1. This well, with depth 5.2 eV and width 3.9 eV, supports two bound states at -4.036 eV and -1.048 eV. When an eight well system A^8 is fabricated, two energy bands of 8-plets exist. One is deep and narrow and centered around -4 eV. The other is less tightly bound and broader, centered near -1 eV.

Suppose now it is desirable to dope this device with an impurity, represented by a potential B, which introduces two levels between these bands. One should be slightly below the upper band, the other slightly above the lower band. Since the upper band is broader than the lower band, we try to design the impurity well to produce an impurity state at -3.5 eV, slightly above the lower band near -4.0 eV, and another state slightly below the broader upper band, at around -2.5 eV. These two levels need not be the two lowest states of this impurity.

The basic idea is to search for a potential with bound states at -3.5 and -2.5 eV with the expectation that the corresponding levels in the multiwell potential won't be too much different. We can design the impurity potential by trial and error or by looking up a table of impurity well energy levels. Both methods are in use. We will use inspired (educated) guesswork. Assume the impurity levels at -3.5 and -2.5 eV are the second and third levels of a square well potential. We can use the resonance

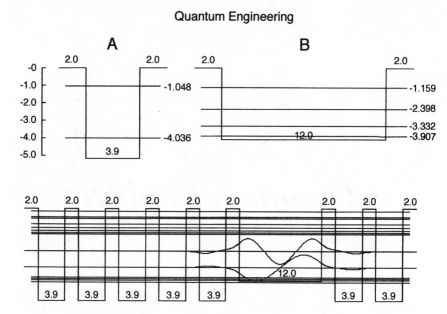

Fig. 34.1 Well A supports two bound states while B supports four. Two of the bound states from well B appear as impurity levels between the two bands of A states in the potential A^5BA^2. Wavefunctions for the impurity states show localization near the B well and parentage of the states.

condition (23.2) to estimate the parameters of this potential:

$$2\sqrt{2m(E_n - V)}L/\hbar = 2\pi n , \qquad (34.1)$$

with $E_2 = -3.5$ eV and $E_3 = -2.5$ eV. Solving these two equations provides an estimate for the well parameters: $V \sim -4.3$ eV, $L \sim 14$ Å. These values can be used as starting points for more refined well designs.

The potential (B) with depth 4.1 eV and width 12 Å supports four bound states (Fig. 34.1). The middle two at $E_2 = -3.332$ eV and $E_3 = -2.398$ eV have the desired properties of lying between the bands produced by the potential A^N. At the bottom of Fig. 34.1 we show a potential in which one of the "atoms" in the A^8 device has been replaced by a B "atom." The energy bands for this impurity-doped device A^5BA^2 show 7-plets that arise from the A well, which appears seven times. Two impurity levels arising from the B impurity fall nicely between the two bands, at -3.892 eV and -2.404 eV. They don't appear exactly where expected (the lower level is 0.4 eV lower than hoped for), but they satisfy the original criteria. One level is slightly above the lower band, the other level is slightly below the upper band. The remaining two levels of the B impurity are mixed into the lower and upper bands.

We have also plotted the wavefunctions of these two impurity states in Fig. 34.1. These wavefunctions show that the electron is localized near the B impurity site, with small probability of occurring in the adjacent A sites. The probability for being in the next nearest neighbor well is already too small to be seen at this level of resolution and decreases rapidly with distance from the impurity site.

These wavefunctions also confirm that the impurity levels arise from the second and third levels in the B impurity because they have one and two nodes within the B well. They have many ($= 7 + 1 - 1$) other nodes outside the B well that cannot be seen because the amplitudes are so small. The total number of nodes identifies how many states exist below each level in this A^5BA^2 potential. The number of nodes in well B identifies the parent state in the isolated B well.

35

Variations on a Theme

We have seen that when a potential consists of two different wells separated by a small distance, the eigenvalues and eigenfunctions of the double-well potential can easily be inferred from those of the isolated wells. We have also seen that the properties of a set of N-identical wells can easily be predicted from the eigenvalues and eigenfunctions of a simple isolated well.

We now inquire whether it is possible to determine the behavior of a multiple well potential composed of two different potentials, A and B, arranged in any order. In Fig. 35.1 we show two isolated wells: A with depth 5.0 eV and width 8.0 Å, and B with depth 4.0 eV and width 5.0 Å. The wells are surrounded by shoulders at 0.0 eV of width 1.0 and 1.5 Å, respectively. Since well A is deeper and wider than well B, it supports more bound states. In Table 35.1 we list the energy eigenvalues for the isolated wells A and B as well as the scattering resonances below 5.0 eV. We also list these properties for various multiwell potentials built from A and B arranged in various orders. As in Part II, chapter 19, we do not discuss properties of the mirror image wells (e.g., BBA $\sim$ ABB), as they are equivalent by time reversal invariance. We also note that $N + 1$ barriers (Part II, chapter 19) produce N wells, so that Table 35.1 contains more information than Table 19.1 in Part II.

The principle results for this class of potentials can be seen from the table and can be guessed from results of previous calculations. A multiwell potential containing N_A copies of well A and N_B copies of well B will have as eigenvalues N_A-plets of the A well eigenvalues occurring in bands and N_B-plets of the B well eigenvalues, also in bands. The deeper the bands, the narrower. The location of these bands is generally close to the location of the eigenvalues of the isolated wells. In addition,

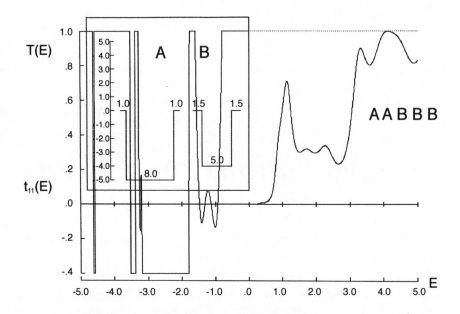

Fig. 35.1   Plot of $t_{11}(E)$ (-5.0 eV $< E < 0.0$ eV) and $|1/t_{11}(E)|^2$ (0.0 eV $< E < 5.0$ eV) for the potential A^2B^3. The isolated potentials are shown in the inset.

the transmission resonances will also occur in N_A- and N_B-plets. However, some of the resonances may overlap so strongly that they cannot be resolved.

What difference does the order of the wells in the N-well potential have on the eigenvalue distribution? To see this, we can study the series of potentials AB^mA B^{N-m-2} with N wells. The A levels occur in doublets, the B levels in $(N-2)$-plets. When the A wells are adjacent, the splitting is

$$\Delta E \sim \exp - \int_a^b \kappa(x)dx \sim \exp[-\sqrt{2m(V - E_n)/\hbar^2}(2.0)] \ .$$

However, as the separation between the A levels increases ($m > 0$), the splitting becomes smaller. For the A level at -4.6 eV which lies below the bottom of the B well, the splitting approaches $\exp[-\sqrt{2m(4.6)/\hbar^2}(2.0) - \sqrt{2m(4.6 - 4.0)/\hbar^2}(8.0)]$. For the A levels above the base of the B potential the splitting also decreases with increasing m because of interference effects in the B wells. In short, the more separated the A wells, the smaller the splitting.

It is also possible to study the effect of order on the wavefunctions. We present only a brief description here. In an isolated well (see B in A^5BA^2, Fig. 35.1), the B well wavefunctions are basically localized to the B well, with little overlap to the

Table 35.1 Bound state eigenvalues ($E < 0$) and resonance transmission peaks ($E > 0$) for multiwell potentials

Potential	Eigenvalues and Peaks
A	$-4.61, -3.45, -1.64;\ 0.28, 4.39$
B	$-3.23, -1.19;\ 2.01$
AA	$(-4.62, -4.59), (-3.52, -3.38), (-1.79, -1.46);\ (0.28, 0.81), (3.81, 4.39)$
AB	$-4.61, -3.45, -3.23, -1.67, -1.13;\ 0.50, 2.02, 3.93$
BB	$(-3.26, -3.21), (-1.30, -1.03);\ (1.51, 2.01), 4.82$
AAA	$(-4.63, -4.61, -4, 59), (-3.54, -3.45, -3, 36), (-1.85, -1.63, -1.37);$ $(0.28, 0.50, 1.15), (3.32, 4.34, 4.39)$
AAB	$(-4.62, -4.59), (-3.52, -3.39), -3.23, (-1.80, -1, 49), -1.11;$ $(0.37, 0.91), 1.91, (3.46, 4.26)$
ABA	$-4.61(2), (-3.46, -3.45), -3.22, (-1.70, -1.63) - 1.07;\ 0.47, 1.87, 3.82$
ABB	$-4.61, -3.45, (-3.26, -3.21), -1.67, -1.27, -1.00;\ 1.40, 2.14, 3.82, 4.71$
BAB	$-4.61, -3.46, (-3.23, -3.22), -1.70, -1.18, -1.07;\ 0.55, 1.97, 3.64$
BBB	$(-3.27, -3.23, -3.20), (-1.35, -1.18, -0.96);\ (1.16, 1.93, 2.01), 4.18$
AAAA	$(-4, 63, -4.61, -4.59, -4.58), (-3.56, -3.49, -3.41, -3, 34), (-1.88,$ $-1.73, -1.52, -1, 32);\ (0.28, 0.38, 0.81, 1.33), 3.10, 4.39, 4.61$
AAAB	$(-4, 63, -4.61, -4.59), (-3.55, -3.45, -3, 36), -3.23, (-1.85, -1.64,$ $-1, 38), -1.11;\ 0.62, 1.18, 3.19, 3.91$
AABA	$(-4, 62, -4.61, -4.59), (-3.52, -3.45, -3, 39), -3.22, (-1.80, -1.67,$ $-1, 48), -1.05;\ 0.41, 0.90, 1.89, 3.30, 3.92$
AABB	$(-4, 62, -4.59), (-3.52, -3, 39), (-3.26, -3.21), (-1.80, -1, 49),$ $(-1.26, -1.00);\ 0.93, 1.33, 2.17, 3.36, 4.70$
ABAB	$-4.61(2), (-3.46, -3, 45), (-3.23, -3.21), (-1.71, -1, 64), (-1.15,$ $-1.03);\ 0.57, 1.85, 3.43$
ABBA	$-4.61(2), -3.46(2), (-3.25, -3.20), (-1.68, -1, 66), (-1.24, -0.97);$ $0.49, 1.34, 2.17, 3.72, 4.66$
BAAB	$(-4, 62, -4.59), (-3.52, -3, 39), (-3.23, -3.22), (-1.80, -1, 51),$ $(-1.14, -1.09);\ 0.36, 0.95, 1.96, 3.30, 4.13$
ABBB	$-4.61, -3.45, (-3.27, -3, 23, -3.20), -1.67, (-1.33, -1.14, -0.94);$ $1.13, 1.79, 2.25, 4.22$
BABB	$-4.61, -3.46, (-3.26, -3, 23, -3.20), -1.70, (-1.28, -1.13, -0.99);$ $0.55, 2.08, 3.51, 4.57$
BBBB	$(-3.28, -3, 25, -3.22, -3.19), (-1.37, -1.25, -1.09, -0.92);$ $1.02, 1.51, 2.01, 2.16, 3.88, 4.82$

adjacent A wells. In addition, two of the A eigenfunctions in that potential look like symmetrized and antisymmetrized wavefunctions in a double-well A^2 potential (Fig. 31.1). These have little leakage into the neighboring B well and less into the A^5 part of the potential on the far side of the B well. Conversely, five of the eigenstates with A-well parentage are localized in the A^5 part of the potential, with small leakage into the BA^2 part of the potential.

This behavior tends to be the case in general and becomes more pronounced the deeper the levels. Eigenstates in potentials $A^{m_1} B^{n_1} A^{m_2} B^{n_2} \ldots$ tend to be localized to the m_j-tuple A^{m_j} of the potential and occur in an m_j subband with degeneracy m_j. These states have little leakage into the adjacent B wells $B^{n_j - 1}$, B^{n_j}. Further, these eigenstates are very similar to the symmetrized eigenstates studied in Chapter 31. Similar statements can be given for eigenstates with B-well parentage.

In Fig. 35.1 we show a plot of the transfer matrix element $t_{11}(E)$, whose zero crossings locate the energy eigenvalues. We extend this search above 0.0 eV by plotting the transmission probability, $|1/t_{11}(E)|^2$. The calculation shown is carried out for the potential $A^2 B^3$. The double crossings of $t_{11}(E)$ for the A doublets occur in the neighborhood of -4.6, -3.5, and -1.6 eV. Triple crossings for the B-well triplets occur near -3.2 and -1.2 eV. Resonance peaks are more perturbed from their isolated well positions than are the bound states. Peaks generally do not have maximum value at $T(E) = +1$.

The locations of all bound states and peaks for potentials containing up to four wells are summarized in Table 35.1. The calculations were carried out with an energy resolution of 0.01 eV. The multiplicity of unresolved bound states is shown in parentheses. All bound states have negative energy; resonance peaks have positive energy.

Problem. Identify the parentage of all states listed in Table 35.1.

36

The Sine Transform

The method that we are exploiting to solve Schrödinger's equation in one dimension works well for potentials that can be well represented by a piecewise constant potential. Some potentials cannot be well approximated this way. These include

1. Long-range potentials:

$$x < 0 \qquad 0 \leq x$$
$$V(x) = 0 \quad V(x) = \tfrac{-1.0}{x-x_0} \quad x_0 = -10.0 \,. \qquad (36.1)$$

2. Very deep potentials:

$$x \leq 0 \qquad 0 \leq x \leq 1.0 \qquad 1 \leq x$$
$$V(x) = -1.0 \quad V(x) = \tfrac{-1.0}{x} \quad V(x) = -1.0 \,. \qquad (36.2)$$

3. Long-range and infinitely deep potentials:

$$x \leq 0 \qquad 0 \leq x$$
$$V(x) = 0.0 \quad V(x) = \tfrac{-1.0}{x} \,. \qquad (36.3)$$

There are two simple ways to study such potentials. The first is useful for the study of both bound and excited states. It is a natural extension of the methods that we have used so far. The second is applicable only to bound states. It is very elegant and exploits our understanding of the structure of bound state wavefunctions.

In the first case, we assume a very fine mesh for the evaluation of wavefunctions, with breakpoints for a piecewise approximation at x_j, $j = \text{``}-\infty\text{''}, \ldots, -2,$

$-1, 0, 1, 2, \ldots,$ "$+\infty$." We choose as basis states in region j, $x_{j-1} \leq x \leq x_j$:

$$
\begin{array}{lll}
E > V & \cos k(x - x_{j-1}) & \sin k(x - x_{j-1})/k , \\
E = V & 1 & (x - x_{j-1}) , \\
E > V & \cosh \kappa(x - x_{j-1}) & \sinh \kappa(x - x_{j-1})/\kappa .
\end{array}
\tag{36.4}
$$

Then matching boundary conditions between regions j and $j+1$ at x_j leads to equations for the amplitudes $\begin{bmatrix} A \\ B \end{bmatrix}_j$ and $\begin{bmatrix} A \\ B \end{bmatrix}_{j+1}$, which are, for $E > V$,

$$
\begin{array}{llll}
A_j \cos k(x_j - x_{j-1}) & + & B_j \sin k(x_j - x_{j-1})/k & = & A_{j+1} , \\
-k A_j \sin k(x_j - x_{j-1}) & + & B_j \cos k(x_j - x_{j-1}) & = & B_{j+1} .
\end{array}
\tag{36.5}
$$

Now we do what physicists normally do. We write $A_j = A(x_j)$, $A_{j+1} = A(x_{j+1})$, $x_{j+1} - x_j = dx = x_j - x_{j-1}$, and rewrite these equations in the limit dx small:

$$
\begin{array}{llll}
A(x) & + & B(x)dx & = & A(x + dx) , \\
-k^2 A(x)dx & + & B(x) & = & B(x + dx) .
\end{array}
\tag{36.6}
$$

From here it is a simple step to the coupled pair of ordinary differential equations

$$
\frac{dA}{dx} = B ,
$$
$$
\frac{dB}{dx} = -k^2 A = -\frac{2m}{\hbar^2}(E - V(x)) A .
\tag{36.7}
$$

This last equation is valid for $E > V$, $E = V$, and $E < V$.

We could have guessed this result from the outset. For dx very small, the wavefunction in any region is $A \cos kdx + B \sin kdx/k \to A$. By defining $\psi = A$ and $d\psi/dx = B$, we find Schrödinger's equation

$$
\frac{dA}{dx} = B ,
$$
$$
\frac{dB}{dx} = \frac{d^2\psi}{dx^2} = -\frac{2m}{\hbar^2}(E - V(x))\psi = -\frac{2m}{\hbar^2}(E - V(x)) A .
\tag{36.8}
$$

What has been done here is simply to reduce a single second-order ordinary differential equation to two coupled first-order differential equations, in the spirit of the transition from Lagrangian to Hamiltonian mechanics.

These equations are integrated as previously described. We illustrate for a bound state. The wavefunction must vanish in the asymptotic limit $x \to -\infty$. In the asymptotic regime where $V_{L,R}$ are constant and $E < V_L, V_R$, initial conditions for integration are

$$
\begin{aligned}
\psi(x) &= A\left(\frac{e^{\kappa x} + e^{-\kappa x}}{2}\right) + \frac{B}{\kappa}\left(\frac{e^{\kappa x} - e^{-\kappa x}}{2}\right) \\
&= \frac{1}{2}\left(A + \frac{B}{\kappa}\right)e^{\kappa x} + \frac{1}{2}\left(A - \frac{B}{\kappa}\right)e^{-\kappa x} .
\end{aligned}
\tag{36.9}
$$

Boundary conditions are

$$
A + \frac{B}{\kappa} \qquad A - \frac{B}{\kappa}
$$

	$A + \frac{B}{\kappa}$	$A - \frac{B}{\kappa}$	
$-\infty \leftarrow x$	$\neq 0$	0	(36.10)
$x \rightarrow +\infty$	0	$\neq 0$.	

Initial conditions $A - B/\kappa = 0$, $A + B/\kappa = 1$ are chosen at $x = -\infty$, and the integral is carried to $+\infty$. Those states for which $A + B/\kappa = 0$ as $x \rightarrow +\infty$ are eigenstates.

Problem. If eigenstates are chosen as $\cos kx$, $\sin kx$ $(E > V)$ and $\cosh \kappa x$, $\sinh \kappa x$ $(E < V)$ instead of $\sin kx/k$, $\sinh \kappa x/\kappa$ as in (36.4), then show that the coupled equations take the trigonometric and hyperbolic forms

$$
\begin{bmatrix} \dfrac{d}{dx} & -k \\ +k & \dfrac{d}{dx} \end{bmatrix} \begin{bmatrix} A \\ B \end{bmatrix} = 0 \qquad E > V,
$$

$$(36.11)$$

$$
\begin{bmatrix} \dfrac{d}{dx} & -\kappa \\ -\kappa & \dfrac{d}{dx} \end{bmatrix} \begin{bmatrix} A \\ B \end{bmatrix} = 0 \qquad E < V.
$$

We now turn to a second method for computing bound states of a potential. This is based on a deep understanding of the structure of eigenfunctions for bound states.

The first observation is that they "wiggle" a lot. The higher the energy, the more they wiggle. To be more precise, they look like sine functions. In fact, for piecewise constant potentials they are sine functions $(E > V)$, possibly displaced in phase: $\sin(kx + \phi_0)$. If the potential is not constant, one might expect wavefunctions could be written in the form $\sin(k(x)x + \phi_0)$, with a position-dependent momentum $p(x) = \hbar k(x)$. The second observation is that the amplitude of the wavefunction is not constant and in fact may grow or decrease rapidly in classically forbidden regions $E < V$ when the wavefunction is not being very sine-like. Therefore one might expect that

$$
\psi(x) = A(x) \sin \phi(x) \tag{36.12}
$$

could provide a reasonable representation of a wavefunction for a bound state in any kind of potential.

This is not quite the ansatz (functional form) we would like to take. We would like to ensure that all wavefunction nodes occur due to the increase of the phase $\phi(x)$, and not because of zeros of the amplitude. To enforce the condition that the amplitude is always positive, we write the amplitude as the exponential of a real function, so that

$$
\psi(x) = e^{S(x)} \sin \phi(x) . \tag{36.13}
$$

This is called the "sine transform" (or "Prufer transform") of the wavefunction.

What equation (36.13) does for us is at first sight ridiculous. As if life weren't complicated enough with one unknown function $\psi(x)$, we now have two! While there is some hope of computing $\psi(x)$ once the boundary conditions are established and satisfied, we have a less "rigid" system now with two unknown functions. However, we can use this extra degree of freedom as an asset rather than treat it as a liability. That is, we can impose an extra condition on these two functions. Such a condition could (will) impose a unique relation among the three real functions $\psi(x), S(x)$, and $\phi(x)$.

The condition we impose is

$$\frac{d\psi}{dx} = \psi' = e^{S(x)} \cos \phi(x) \, . \tag{36.14}$$

The equations that the two functions $S(x), \phi(x)$ satisfy are derived as follows:

$$\frac{d\psi}{dx} \overset{\text{by calculus}}{=} e^S \cos \phi \, \phi' + e^S \sin \phi \, S' \overset{\text{by (36.14)}}{=} e^S \cos \phi \, ,$$

$$\frac{d^2\psi}{dx^2} \overset{\text{by (36.14)}}{=} -e^S \sin \phi \, \phi' + e^S \cos \phi \, S' \overset{\text{by (1.6)}}{=} -\frac{2m}{\hbar^2}(E - V)e^S \sin \phi. \tag{36.15}$$

The factor e^S is common to all terms in both equations. Simplification leads to

$$\begin{bmatrix} \cos \phi & \sin \phi \\ -\sin \phi & \cos \phi \end{bmatrix} \begin{bmatrix} \phi' \\ S' \end{bmatrix} = \begin{bmatrix} \cos \phi \\ -\frac{2m}{\hbar^2}(E - V)\sin \phi \end{bmatrix} . \tag{36.16}$$

This is a simple equation to solve. The matrix on the left is an orthogonal matrix. Its inverse is its transpose

$$\begin{bmatrix} d\phi/dx \\ dS/dx \end{bmatrix} = \begin{bmatrix} \cos \phi & -\sin \phi \\ \sin \phi & \cos \phi \end{bmatrix} \begin{bmatrix} \cos \phi \\ -\frac{2m}{\hbar^2}(E - V)\sin \phi \end{bmatrix}$$

$$= \begin{bmatrix} \cos^2 \phi + \frac{2m}{\hbar^2}(E - V)\sin^2 \phi \\ \left(1 - \frac{2m}{\hbar^2}(E - V)\right)\sin \phi \cos \phi \end{bmatrix} . \tag{36.17}$$

The remarkable feature of these two equations is that the equation for $\phi(x)$ is independent of the amplitude S (or e^S):

$$\frac{d\phi}{dx} = \cos^2 \phi + \frac{2m}{\hbar^2}(E - V)\sin^2 \phi \, . \tag{36.18}$$

Further, this equation has remarkable properties. The wavefunction must vanish in the asymptotic regions $x \to \pm\infty$. The wavefunction must vanish through the sine term, not the amplitude. Therefore, ϕ is an integer multiple of π in both asymptotic regions. In particular, the difference $\phi(+\infty) - \phi(-\infty)$ is an integer multiple of π. By setting the initial value $\phi(x \to -\infty) = 0$, we find

$$\int_{-\infty}^{+\infty} \frac{d\phi}{dx} dx = \phi(x \to +\infty) - \phi(x \to -\infty) = n\pi \, . \tag{36.19}$$

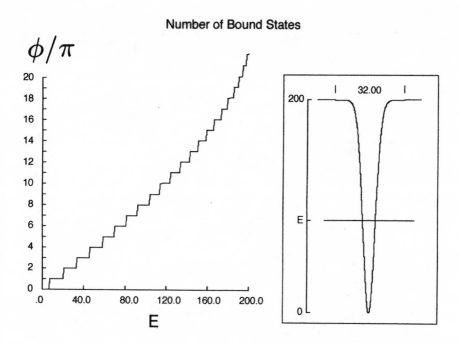

Fig. 36.1 Plot of $\Delta\phi(E)/\pi$ versus E for Gaussian potential shown in inset. This function is integer valued, with the integer being the number of bound states that exists below energy E.

A theorem states that n is the number of bound states with energy E_i less than or equal to E, the energy for which the integral is carried out.

As a result, if we integrate equation (36.18) and plot the result as a function of increasing energy, we should see a series of discontinuous jumps as the energy E exceeds the eigenvalues of successive bound states.

In Fig. 36.1 we present a plot of ϕ/π as a function of E for the Gaussian potential

$$V(x) = V_0[1 - e^{-(x/L)^2}], \qquad (36.20)$$

with $V_0 = 200.0$ eV, $L = 4.0$ Å. This result has plateaus at integer values and is in some ways similar to Fig. 23.2, which plots the number of bound states as a function of Action for the same potential.

Wavefunctions themselves can be constructed in this representation without difficulty. The energy, E_i, at which a jump in $\Delta\phi$ takes place is determined to desired accuracy by any convenient numerical means. Then the coupled first-order equations (36.17) for ϕ and S are simultaneously integrated with initial conditions $\phi(x \to -\infty) = 0$, $S(x \to -\infty) = 0$. The solutions provide the wavefunctions through (36.13).

Part IV

Periodic Potentials

37

Boundary Conditions

Regular crystalline solids consist of atoms, or groups of atoms, packed together in a regular lattice array. In the real world crystals are three dimensional and therefore their treatment falls outside the scope of our work. However, a great deal of information about crystals can be obtained by looking at them along only one spatial dimension and treating them as regular one-dimensional arrays.

Even a very small crystal consists of very many (N) identical units (unit cells) repeated in a regular array. To determine the properties of a one-dimensional crystal (electron energy eigenvalue spectrum, transmission spectrum, wavefunctions), appropriate boundary conditions must be chosen. There are three practical possibilities:

1. **Scattering boundary conditions:** The asymptotic potentials (V_L, V_R) on the left and right are less than any of the internal potential energies of the unit cell.

2. **Bound state boundary conditions:** The asymptotic potentials (V_L, V_R) are greater than any of the other energies under consideration, so only bound states occur.

3. **Periodic boundary conditions:** The wavefunction in cell i is identical to the wavefunction in cell $i + N$ (any i).

As might be expected (and hoped), the physical results are insensitive to the boundary conditions chosen and become identical as N (number of unit cells in the lattice) grows large. For this reason, we impose the boundary conditions that are simplest to implement. Periodic boundary conditions are *far* easier to implement than either scattering or bound state boundary conditions for N large ($N > 3$).

In Fig. 37.1 we illustrate the idea of periodic boundary conditions. On the left in Fig. 37.1 we show a piece of a one dimensional periodic potential which extends off to infinity in both directions. To impose periodic boundary conditions we identify the 0th unit cell with the Nth unit cell, the 1st with the $(N+1)$st, and so on. Then the potential consists of N copies of the potential associated with a unit cell, and these copies are arranged at equal intervals around the circumference of a circle, as shown on the right in this figure.

PERIODIC POTENTIALS

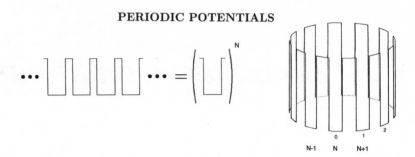

Fig. 37.1 *Left*, A periodic one-dimensional potential is represented by the potential for a single unit cell that repeats to infinity on the left and right. *Right*, When the ith cell is identified with the $(i+N)$th cell, the potential can be visualized as consisting of N identical unit cell potentials equally distributed around the circumference of a circle.

If (A_i, B_i) are the two independent amplitudes in a certain part of cell i, and T is the transfer matrix for a single unit cell, then

$$\begin{pmatrix} A \\ B \end{pmatrix}_i = T \begin{pmatrix} A \\ B \end{pmatrix}_{i+1} . \tag{37.1}$$

This is true for each unit cell, so that

$$\begin{pmatrix} A \\ B \end{pmatrix}_i = T \begin{pmatrix} A \\ B \end{pmatrix}_{i+1} = T^2 \begin{pmatrix} A \\ B \end{pmatrix}_{i+2} = \cdots = T^N \begin{pmatrix} A \\ B \end{pmatrix}_{i+N} . \tag{37.2}$$

Under the assumption of periodic boundary conditions, $\begin{pmatrix} A \\ B \end{pmatrix}_i = \begin{pmatrix} A \\ B \end{pmatrix}_{i+N}$ for all i, so that

$$T^N = I_2 , \tag{37.3}$$

where I_2 is the unit 2×2 matrix.

In order to progress, we introduce a nonsingular 2×2 matrix, S, and carry out a similarity transformation. Then

$$ST^N S^{-1} = ST \underbrace{S^{-1}ST}_{} \underbrace{S^{-1}}_{} \cdots \underbrace{S}_{} TS^{-1} = (STS^{-1})^N = I_2 . \tag{37.4}$$

If we now choose S to diagonalize T, we find

$$STS^{-1} = \begin{bmatrix} \lambda_1 & 0 \\ 0 & \lambda_2 \end{bmatrix} . \tag{37.5}$$

As a further simplification

$$\det(STS^{-1}) = \det(S)\det(T)\,[\det(S)]^{-1} = \det(T) . \tag{37.6}$$

Further, the transfer matrix across a unit cell satisfies $\det(T) = +1$, so that $\lambda_1\lambda_2 = +1$, or

$$STS^{-1} = \begin{bmatrix} \lambda & 0 \\ 0 & \lambda^{-1} \end{bmatrix} . \tag{37.7}$$

The quantization condition (37.3) then simplifies to

$$(STS^{-1})^N = \begin{bmatrix} \lambda^N & 0 \\ 0 & \lambda^{-N} \end{bmatrix} = \begin{bmatrix} +1 & 0 \\ 0 & +1 \end{bmatrix} . \tag{37.8}$$

In computing the eigenvalues of T, two cases arise:

Case 1: λ Real, $|\lambda| \neq +1$. In this case, $\lambda^N \neq +1$, so the quantization condition cannot be satisfied.

Case 2: λ Complex. In this case $|\lambda| = +1$ and $\lambda = e^{i\phi} = \cos\phi + i\sin\phi$. Then $\lambda^N = e^{iN\phi}$ and λ^N can equal $+1$ when λ is an Nth root of $+1$:

$$\lambda = e^{i2\pi k/N} , \tag{37.9}$$

where k is an integer. The N distinct roots are given explicitly by

$$
\begin{aligned}
k \quad &= \quad 0, 1, 2, \ldots, N-1 \\
\text{or} \quad &\quad\quad 1, 2, 3, \ldots, N \\
\text{or} \quad &\quad\quad -N/2, \ldots, -1, 0, +1, \ldots, +N/2 .
\end{aligned}
\tag{37.10}
$$

We now compute the eigenvalues in terms of the matrix elements of the transfer matrix for the unit cell. The secular equation for the eigenvalues is determined from

$$\det\left[\begin{pmatrix} t_{11} & t_{12} \\ t_{21} & t_{22} \end{pmatrix} - \lambda I_2 \right] = (t_{11} - \lambda)(t_{22} - \lambda) - t_{12}t_{21} = 0 , \tag{37.11}$$

$$
\begin{aligned}
\lambda^2 \quad - \quad (t_{11} + t_{22})\lambda \quad + \quad (t_{11}t_{22} - t_{12}t_{21}) \quad &= \quad 0 , \\
\lambda^2 \quad - \quad \operatorname{tr}(T)\lambda \quad + \quad \det(T) \quad &= \quad 0 .
\end{aligned}
\tag{37.12}
$$

Here $\operatorname{tr}(T)$ and $\det(T)$ are the trace and determinant of the transfer matrix T. Since $\det(T) = +1$, the solutions are

$$\lambda_{\pm} = \frac{1}{2}\operatorname{tr}(T) \pm \sqrt{\left(\frac{1}{2}\operatorname{tr}(T)\right)^2 - 1} . \tag{37.13}$$

It is convenient to remove the factor of -1 from under the radical:

$$\lambda_{\pm} = \frac{1}{2}\text{tr}(T) \pm i\sqrt{1 - \left(\frac{1}{2}\text{tr}(T)\right)^2}. \tag{37.14}$$

When $\left|\frac{1}{2}\text{tr}(T)\right| \leq 1$ it is convenient to define an angle ϕ by

$$
\begin{aligned}
\cos\phi &= \tfrac{1}{2}\text{tr}(T) \\
\sin\phi &= \pm\sqrt{1 - \left(\tfrac{1}{2}\text{tr}(T)\right)^2} = \pm\sqrt{1 - \cos^2\phi}.
\end{aligned}
\tag{37.15}
$$

In this case the eigenvalues of the transfer matrix $T(E)$ are

$$\lambda_{\pm}(E) = e^{\pm i\phi(E)}, \tag{37.16}$$

where $\cos\phi(E) = \frac{1}{2}\text{tr}(T(E))$ and we have shown in (37.16) that the angle $\phi(E)$ and eigenvalues $\lambda_{\pm}(E)$ are explicitly dependent on the energy of an electron in the periodic potential.

The quantization condition $\cos\phi(E) = \frac{1}{2}\text{tr}(T(E))$ and $\phi = 2\pi k/N$ does not uniquely define the phase angle ϕ. Rather, this condition defines ϕ

- up to sign

- up to integer multiples of 2π radians.

We will use this lack of uniqueness to provide a very useful relation between the eigenvalues $\lambda_{\pm}(E)$, the phase angles $\phi(E)$, the transfer matrix $T(E)$, and the energy, E.

Before discussing simple examples, we describe how the eigenvalues $\lambda_{\pm}(E)$, and therefore the angle $\phi(E)$, behave as a function of increasing energy. This behavior is illustrated in Fig. 37.2. For very low energies, the two eigenvalues are real and positive: $0 < \lambda_{-} < 1 < \lambda_{+}$. The phase angle is undefined since $\left|\frac{1}{2}\text{tr}(T)\right| > 1$. As the energy increases to the lowest allowed value, the two eigenvalues approach each other along the positive real axis. They "collide" at $\lambda_{\pm} = +1$ and scatter off each other onto the circumference of the unit circle. As the energy increases, they wind around the circumference of the unit circle until they reach $\lambda_{\pm} = -1$. As they do so, one of the phase angles increases from 0 to π radians, while the other, the negative of the positive phase angle, decreases from 0 to $-\pi$ radians.

When the phase angles reach $\pm\pi$ and the eigenvalues reach $\lambda_{\pm} = -1$, the eigenvalues scatter off each other onto the negative real axis and obey the inequality $\lambda_{-} < -1 < \lambda_{+} < 0$. As the energy continues to increase, they stop separating from each other, and begin to approach each other, until they collide again at $\lambda_{\pm} = -1$. Between the two collisions at $\lambda_{\pm} = -1$, no eigenstates are allowed and the phase angle is undefined since $\left|\frac{1}{2}\text{tr}(T)\right| > 1$.

After this latest collision at $\lambda_{\pm} = -1$, the eigenvalues scatter onto the circumference of the unit circle. One phase angle ϕ (the "positive" angle) then continues to increase in the range $\pi \leq \phi \leq 2\pi$ until the eigenvalues collide at $\lambda = +1$ and scatter

Eigenvalues of T

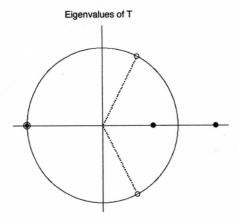

Fig. 37.2   As the energy increases, the eigenvalues move around the circumference of the unit circle in the positive and negative directions, scattering off each other onto the positive or negative real axis at even and odd multiples of π radians. The alternation between complex and real values mirrors the alternation between energy bands and band gaps.

off each other onto the positive real axis. The other phase angle (its negative) decreases from $-\pi$ to -2π in this energy range. The eigenvalues then separate along the real axis and recombine, colliding again at $\lambda_\pm = +1$ and scattering onto the circumference of the unit circle, where the positive phase angle increases once again ($2\pi \leq \phi \leq 3\pi$), and the negative phase angle simultaneously decreases ($-2\pi \geq \phi \geq -3\pi$). This repetitive process continues as the energy E increases.

38

A Simple Example

To illustrate these results, we compute the allowed energy levels for a periodic lattice whose unit cell is shown at the lower right in Fig. 38.1. The unit cell consists of two regions, one of width $L_1 = 8.0$ Å at $V_1 = 0.0$ eV, the second of width $L_2 = 1.0$ Å at $V_2 = 9.0$ eV. We scan the energy between 0.0 and 10.0 eV. The transfer matrices M_1 and M_2 for the two regions of the unit cell for $E < V_2$ are

$$
M_1 = \begin{bmatrix} \cos k_1 L_1 & -\dfrac{1}{k_1} \sin k_1 L_1 \\ k_1 \sin k_1 L_1 & \cos k_1 L_1 \end{bmatrix} , \qquad k_1 = \sqrt{2mE/\hbar^2} . \tag{38.1}
$$

$$
M_2 = \begin{bmatrix} \cosh \kappa L_2 & -\dfrac{1}{\kappa} \sinh \kappa L_2 \\ -\kappa \sinh \kappa L_2 & \cosh \kappa L_2 \end{bmatrix} , \qquad \kappa = \sqrt{2m(V_2 - E)/\hbar^2} . \tag{38.2}
$$

For $E > V_2$ the matrix M_2 must be replaced by a matrix similar to M_1 (with $L_1 \to L_2$ and $k_1 \to k_2 = \sqrt{2m(E - V_2)/\hbar^2}$).

The allowed energies are determined by computing $M_1(E)M_2(E)$ and taking half the sum of the diagonal terms. For $E < V_2$ we find

$$
\frac{1}{2}\mathrm{tr}(M_1 M_2) = \cos kL_1 \cosh \kappa L_2 + \frac{1}{2}\left(+\frac{\kappa}{k} - \frac{k}{\kappa} \right) \sin kL_1 \sinh \kappa L_2 . \tag{38.3}
$$

For $E > V_2$ the result is

$$
\frac{1}{2}\mathrm{tr}(M_1 M_2) = \cos k_1 L_1 \cos k_2 L_2 + \frac{1}{2}\left(-\frac{k_2}{k_1} - \frac{k_1}{k_2} \right) \sin k_1 L_1 \sin k_2 L_2 . \tag{38.4}
$$

Eigenvalues of Periodic Potentials

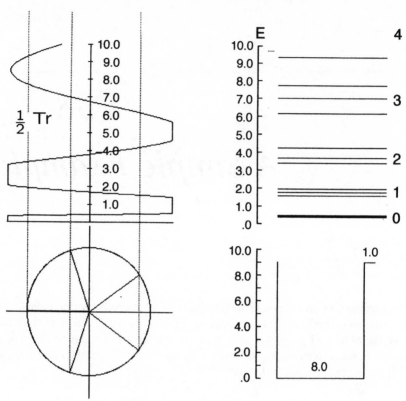

Fig. 38.1 *Lower right*, Potential for a unit cell. *Upper left*, Plot of $\frac{1}{2}\text{tr}(T(E))$ as a function of E, where $T(E)$ is the unit cell transfer matrix. The energy axis is vertical. The value of $\frac{1}{2}\text{tr}(T(E))$ is truncated below -1.5 and above $+1.5$. *Lower Left*, Unit circle. Angles are measured from the heavy line segment. The allowed phase angles $0, \pm 2\pi/5, \pm 4\pi/5$ for a periodic lattice with $N = 5$ are shown. Intersection of vertical dotted lines with curve $\frac{1}{2}\text{tr}(T(E))$ (top left) indicate allowed energy eigenvalues. These have been plotted in an energy level diagram (top right). Each band of allowed energies is labeled by an integer $n = 0, 1, 2, \dots$. Allowed levels are nondegenerate or doubly degenerate, as described in the text.

This function is plotted in Fig. 38.1 (upper left). The plot has been truncated for $\frac{1}{2}\text{tr} < -1.5$ and $\frac{1}{2}\text{tr} > +1.5$ as the interesting part of this plot lies in the range $-1 \leq \frac{1}{2}\text{tr}(T(E)) \leq +1$.

So far, the results are independent of the number of cells in the periodic potential. To show how to compute the allowed energy eigenvalues for a lattice with N cells, we choose a small lattice with $N = 5$. The five allowed phase angles are $k \times 2\pi/5$, with $k = -2, -1, 0, +1, +2$. These five angles are shown on the unit circle in the lower left of Fig. 38.1. They are measured from the fiducial direction (dark line segment). The values for which $\frac{1}{2}\text{tr}(T)$ is equal to $\cos 2\pi k/5$ are shown by the intersections of the three vertical dotted lines through the five angles $0, \pm 2\pi/5, \pm 4\pi/5$ with the function $\frac{1}{2}\text{tr}(T(E))$. Each intersection of the dotted line through the one angle $\phi = 0$ with $\frac{1}{2}\text{tr}(T(E))$ describes one allowed energy eigenvalue. Each intersection of the dotted line through the two angles $\pm 2\pi/5$ describes two allowed energy eigenvalues, one for each angle. The same is true for each intersection of the dotted line through the two angles $\pm 4\pi/5$. These intersections are plotted in an energy-level diagram, shown in the upper right of Fig. 38.1. In this energy-level plot, each band of allowed energy levels is labeled by an integer ($n = 0, 1, 2, \ldots$). Each band contains five allowed energy levels. Two pairs of levels are doubly degenerate and one is nondegenerate. The nondegenerate level is the lowest level in the even ($n = 0, 2, 4, \ldots$) bands and is the highest level within the odd ($n = 1, 3, 5, \ldots$) bands.

The periodic potential with $N = 6$ differs only slightly from the potential with $N = 5$. In this case, each band has six allowed energy eigenvalues. These correspond to the six angles $\phi = 0, \pm 2\pi/6, \pm 4\pi/6$, and π. As a result, within each band, two pairs of levels, corresponding to phase angles $\phi = \pm 2\pi/6, \pm 4\pi/6$ are doubly degenerate, and both the lowest and highest levels within each band, corresponding to phase angles $\phi = 0, \pi$ are nondegenerate.

For large values of N ($N \geq 10$) it is not useful to plot each allowed energy eigenvalue. The reason is that there are so many allowed eigenvalues over a small energy interval that the distinct eigenvalues cannot be resolved. As a result, alternative and more useful methods have been developed to describe the energy dependence of the eigenvalues. One of these methods, the density of states plot, is illustrated in the lower right part of Fig. 38.2. For large N, the number of allowed energy eigenvalues within an energy interval of length ΔE around E is equal to the number of allowed phase angles corresponding to that energy range.

The density of states, $\rho(E)$, describes the number of states in an energy range ΔE containing the energy E. To estimate $\rho(E)$, we compute $\phi(E + \Delta E)$ and subtract $\phi(E)$ from it. The angular difference, $\Delta\phi$, is proportional to the number of states in the energy interval ΔE. Since the number of states in any band of energies is N, we have

$$\frac{dN}{N} = \frac{\rho(E)dE}{N} = \frac{d\phi}{2\pi} , \tag{38.5}$$

$$\rho(E) = \frac{N}{2\pi}\frac{d\phi}{dE} . \tag{38.6}$$

To obtain the density of states plot shown in Fig. 38.2 (lower right), we have computed $\phi(E)$ on a fine mesh of energies, and estimated

$$\rho(E) \sim \phi(E_{i+1}) - \phi(E_i) . \tag{38.7}$$

The singularities that appear at the band edges E_e (the upper and lower energy bounds of a band) behave like $\sim 1/\sqrt{|E - E_e|}$. They are called van Hove singularities and are nonremovable in one-dimensional periodic lattice potentials.

Eigenvalues of Periodic Potentials

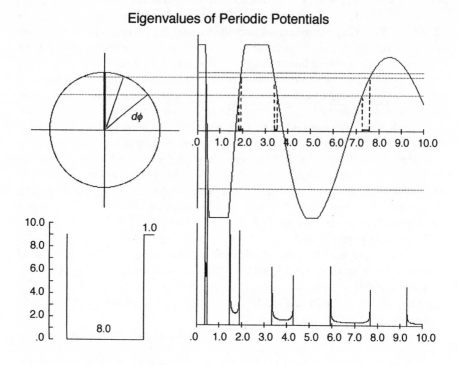

Fig. 38.2   *Lower left,* Potential for a unit cell. *Upper right,* Plot of $\frac{1}{2}\mathrm{tr}(T(E))$ as a function of E, where $T(E)$ is the unit cell transfer matrix. This plot is truncated below -1.5 and above $+1.5$. *Upper left,* Unit circle. The heavy line segment is the fiducial axis, from which angles are measured. The small angular interval $d\phi$ (upper left) describes a small, fixed number of eigenstates, dN, within each energy band. These states are spread over an energy interval, dE, which changes from band to band. The density of states, $\rho(E) = dN/dE$, is proportional to $d\phi/dE$.

Another useful way to describe allowed energy eigenvalues in a periodic potential is by an energy band plot. To construct such a plot, we first plot the allowed energy eigenvalues as a function of the phase angle ϕ. Such a plot is shown on the left of Fig. 38.3. In order to conserve space, each segment of this discontinuous function is translated into the interval $-\pi \le \phi \le +\pi$. The result is shown on the right of this figure. This plot is called an energy band plot. Since it is symmetric about the energy axis $\phi = 0$, often only the positive part of this plot, $0 \le \phi \le +\pi$, is shown. The allowed energies over any value of ϕ correspond to the intersections (shown in Fig. 38.1) of the dotted line through $\cos \phi$ with the plot of $\frac{1}{2}\mathrm{tr}(M_1(E)M_2(E))$.

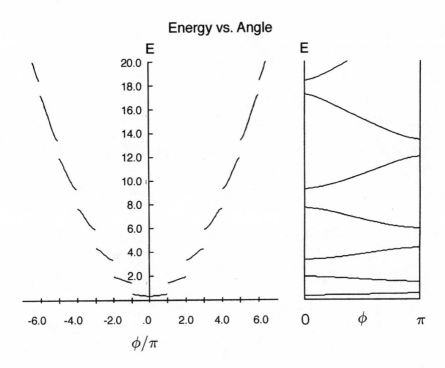

Fig. 38.3 *Left*, Energy as a function of phase angle φ. *Right*, Each segment of this discontinuous function has been moved into the interval $-\pi \leq \phi \leq +\pi$. Only the right half of this figure is shown.

39

Coding and Validation

Equation (37.15) provides a simple algorithm for computing the allowed energy eigenvalues for a particle in a periodic potential from the transfer matrix of the unit cell. However, it is not practical to compute these matrix elements analytically except for the simplest cases (previous chapter). It is therefore useful to carry out such computations numerically. The algorithm for this numerical computation is straightforward:

- Read in the height V_j and width δ_j of each of the pieces comprising a unit cell.

- Choose a value, E, for the particle energy.

- Compute the real 2×2 matrix $M(V_j; \delta_j)$ for each piece of the potential.

- Compute the transfer matrix for the unit cell by multiplying the matrices in the order in which they occur, from left to right: $T(E) = \prod_{j=1}^{k} M(V_j; \delta_j)$.

- Compute $\frac{1}{2}\mathrm{tr}(T(E)) = \frac{1}{2}(t_{11}(E) + t_{22}(E))$.

- For those values of the energy for which $-1 \le \frac{1}{2}\mathrm{tr}(T(E)) \le +1$, compute ϕ.

- Construct density of states plots (Fig. 38.2) or energy band plots (Fig. 38.3) from $\phi(E)$.

In searching for energy eigenvalues by this procedure, this computation must be embedded in a loop that scans over the desired range of energies in which energy bands are to be determined.

Fig. 39.1 shows a simple unit cell in a periodic lattice, along with the band structure and the density of states. Fig. 39.2 is similar to Fig. 39.1, except that the computation

179

Energy Bands and Density of States

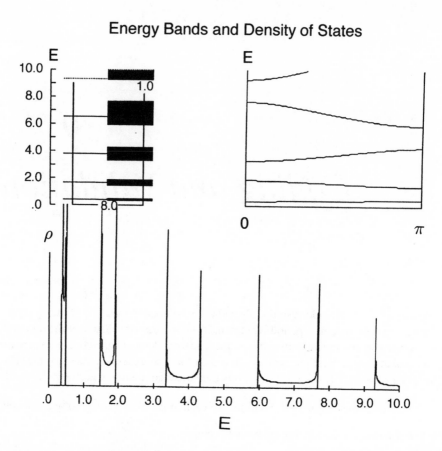

Fig. 39.1 *Upper left*, Binding potential associated with a periodic lattice. The potential has depth 9.0 eV and width 8.0 Å. The four bound state energy eigenvalues (solid lines) and lowest transmission resonance (dotted line) are shown. *Upper right*, Energy bands of the periodic potential. The spacing between the binding potentials is 1.0 Å. The width of the bands is shown in the upper left. *Bottom*, Density of states plot numerically computed for the periodic potential.

has been done numerically for a thirteen-piece approximation to a Gaussian attracting potential. In this calculation the lowest two energy bands have not been resolved by the energy grid used for the scan, nor has the gap between the upper two bands.

Once again we emphasize that numerical codes must be validated by extensive testing before one can rely on the results of these computations. One important way to validate a code is to compare its output with results that are analytically available.

To this end, we numerically reproduced all the results obtained in chapter 38. These numerically obtained results are shown in Fig. 39.1. This figure shows that the energy band and density of states plots obtained numerically are identical to those obtained analytically. In the upper left of Fig. 39.1, we show the bound states (solid lines) and lowest transmission peak (dotted line) for the single well potential with $V = 0$ eV, $L = 8.0$ Å, with asymptotic potentials $V_L = V_R = 9.0$ eV of width $D = 0.5$ Å each. Superposed on these energies, in the right half of this unit cell potential, are the spread of energies that occur in the periodic potential. This figure shows that each isolated state (or resonance) in the potential of a single unit cell is spread out into a band of allowed energies for the periodic potential. Each band of allowed energies includes the energy of the original isolated parent state.

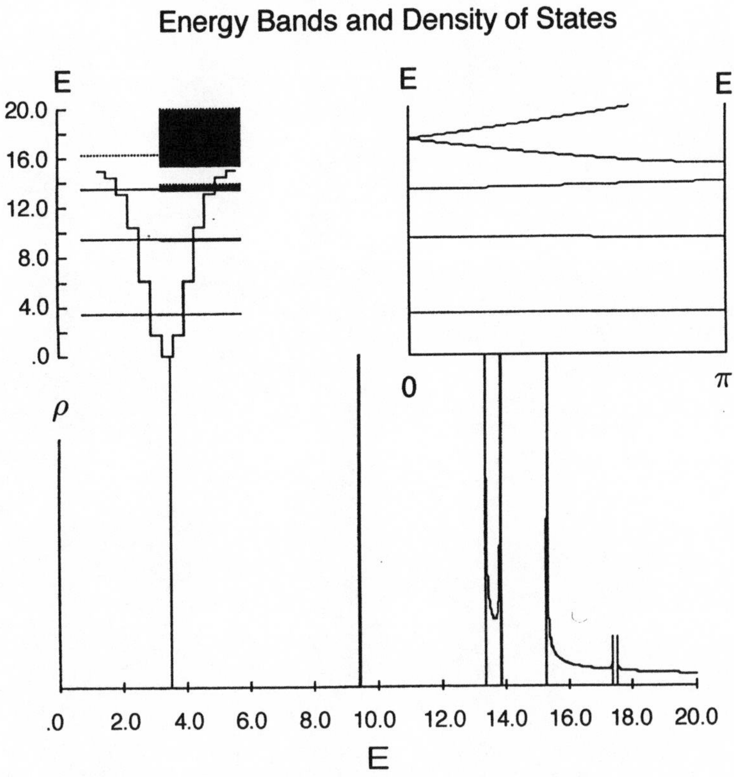

Energy Bands and Density of States

Fig. 39.2 Same as Fig. 39.1, except for a periodic potential whose unit cell is a thirteen-piece approximation to the attracting Gaussian potential well. The lowest two bands ($\sim$3.5 and $\sim$9.5 eV) are hardly spread out and the gap between the fourth and fifth bands ($\sim$17.5 eV) is hardly resolved. The fourth and fifth bands are based on the lowest two transmission resonances.

We should remark here that the cost of computation is totally independent of N, the number of unit cells in the periodic lattice. For this reason, it is useful to invest computer time in an accurate determination of the unit cell transfer matrix, $T(E)$. To speed computation, this can be computed on an energy grid of NN points and stored in an $NN \times 4$ array (each of the four matrix elements is real). In addition, it is often possible to find simple analytic or numerical representations for each of the matrix elements as a function of E, further simplifying the problem of increasing the energy resolution of these scans.

40

Asymptotic Behavior

The plot of energy E versus phase angle ϕ given in Fig. 38.3 looks like a discontinuous approximation to a parabola. In fact, at very high energies it becomes a better and better approximation to the parabola, which relates energy to momentum for a free particle: $E = p^2/2m = (\hbar k)^2/2m$. This suggests that the allowed eigenstates for an electron in a periodic potential are somehow related to free particle states, especially at higher energies.

To make this relation more explicit, in Fig. 40.1 we plot the positive phase angle $\phi(E)$ as a function of $kL = \sqrt{2mE/\hbar^2}L$ for the unit cell shown in Figs. 38.1, 38.2, 38.3, and 39.1. Here $L = L_1 + L_2$ is the width of the unit cell. The steps, which occur at $\phi = j\pi, j = 1, 2\ldots$, indicate regions in which there are no allowed energy eigenvalues (band gaps), so that $\lambda_\pm$ are real. Between steps (e.g., in the allowed bands) the phase increases more or less linearly, except at the band edges where the tangent $(d\phi/d\sqrt{E})$ is vertical. This figure provides three important pieces of information:

1. There is a band gap whenever $\phi = j\pi$, j integer.

2. The gaps become narrower with increasing value of j (or energy).

3. The slope of the plot ϕ/π versus $\sqrt{2mE/\hbar^2}L/\pi$ approaches $+1$ ($\phi \to kL$) rather quickly (except at the band edges).

For this potential, the allowed energies are

$$E(\phi) \sim \frac{\hbar^2}{2m}\left(\frac{\phi}{L}\right)^2, \tag{40.1}$$

Angle and K Vector

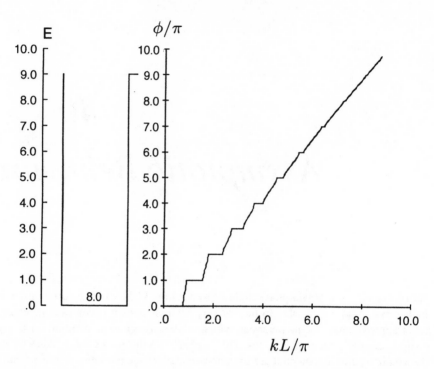

Fig. 40.1 *Left*, Potential in the unit cell. *Right*, Positive phase angle $\phi(E)$ plotted as a function of energy E through its k dependence: kL. Horizontal steps at $\phi = j\pi$ identify band gaps. Nonhorizontal regions indicate energies at which energy bands occur. The slope of this curve approachs ∞ at the band edges. The gaps decrease in width as E increases. As E increases, the slope approaches $+1$ (except at the tiny band gaps), so that $\phi \simeq kL$.

when $\phi > 5\pi$. In short, for energies not too far above the top of the unit cell potential ($V_2 = 9.0$ eV) the electron acts more or less like a free particle with momentum $p = \hbar k$, $k = \phi/L$. The energy eigenstates with $\phi > 0$ represent electrons traveling to the right with momentum $p \sim \hbar k = \hbar\phi/L$. Electrons traveling to the left with momentum $p \sim -|\hbar k| = \hbar\phi/L$ have negative values of ϕ. As the energy E becomes large, or the potential of the unit cell becomes small, the eigenstates of the periodic potential become more like free particle states with momenta $p = \pm\hbar k$ and energy $E = \hbar^2 k^2/2m$.

Band gaps in periodic potentials always include the energies determined by $kL = j\pi$, with j an integer and $k = \sqrt{2mE}/\hbar^2$. The reason is that states that satisfy the condition $kL = (2\pi/\lambda)L = j\pi$ satisfy the Bragg scattering condition $L = j\lambda/2$. This means that intense backscattering interferes with the propagation of even a highly perturbed plane wave. As a result, no allowed eigenstates of a periodic lattice can satisfy this condition.

Fig. 40.1 suggests that band gaps always occur for $\phi = j\pi$, j integer. To illustrate that forbidden energies always exist for periodic potentials, in Fig. 40.2 we plot $\frac{1}{2}\text{tr}(T(E))$ as a function of $kL = \sqrt{2mE/\hbar^2}L$ for the simple unit cell shown in Fig. 40.1. For E large, the maxima and minima occur at regular intervals ($kL = j\pi$). To show that the maximum is always greater than $+1$, and the minimum is always less than -1 for this potential, we proceed as follows. The value of $\frac{1}{2}\text{tr}(T(E))$ is given explicitly when $E > V_2$ by

$$\frac{1}{2}\text{tr}(T(E)) = \cos \phi_1 \cos \phi_2 - \frac{1}{2}\left(\frac{k_1}{k_2} + \frac{k_2}{k_1}\right)\sin \phi_1 \sin \phi_2 , \qquad (40.2)$$

where

$$\begin{aligned}\phi_1 &= k_1 L_1 = \sqrt{2mE/\hbar^2}L_1 \\ \phi_2 &= k_2 L_2 = \sqrt{2m(E - V_2)/\hbar^2}L_2\end{aligned} \qquad , \qquad (40.3)$$

$$\frac{1}{2}\text{tr}(T(E)) = \cos \phi_1 \cos \phi_2 - \frac{1}{2}\left(\sqrt{\frac{1}{1 - (V_2/E)}} + \sqrt{\frac{1 - (V_2/E)}{1}}\right)\sin \phi_1 \sin \phi_2 .$$

$$(40.4)$$

By a standard trigonometric identity (addition formula) for $E \gg V_2$ this simplifies to

$$\frac{1}{2}\text{tr}(T(E)) = \cos (\phi_1 + \phi_2) - \frac{1}{8}\left(\frac{V_2}{E}\right)^2 \sin \phi_1 \sin \phi_2 . \qquad (40.5)$$

The maxima occur close to $\phi_1 + \phi_2 = (\text{integer}) \times 2\pi$. When this occurs, ϕ_1 mod 2π is in the range $0 < \phi_1 \,(\text{mod } 2\pi) < \pi$ and ϕ_2 mod 2π is in the range $\pi < \phi_2 \,(\text{mod } 2\pi) < 2\pi$, or vice versa. In either case, $\sin \phi_1$ and $\sin \phi_2$ have opposite signs, so that the maximum exceeds $+1$.

The minima occur close to $\phi_1 + \phi_2 = (\text{integer}) \times 2\pi + \pi$. In this case, $\sin \phi_1$ and $\sin \phi_2$ are both positive or both negative, so that the minimum is less than -1. As a result, the plot shown in Fig. 40.2 always crosses the limit $+1$ at a maximum and the limit -1 at a minimum, even though these crossings cannot be resolved at the scale of this plot.

This argument shows that band gaps exist for the potential shown in Fig. 40.1, independent of energy range. This argument also shows that these gaps occur around energy $E_j \sim (\hbar j\pi)^2/2mL^2$, where j is an integer.

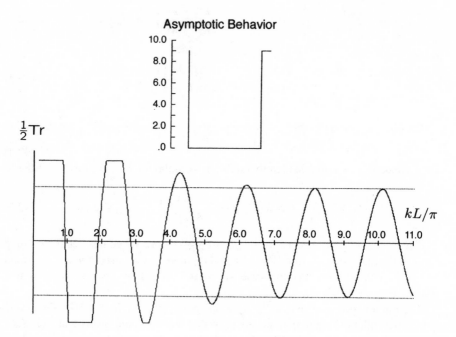

Fig. 40.2 Plot of $\frac{1}{2}\mathrm{tr}(T(E))$ as a function of kL. Asymptotically this function approaches a cosine, with all maxima occurring near even integer multiples of π with values $1 + \epsilon$ and all minima occurring near odd integer multiples of π with values of $-1 - \epsilon$, ϵ small and positive in both cases.

41

Relation among Boundary Conditions

There is a close relationship between the properties of binding potentials, scattering potentials, and periodic potentials that are composed of the same basic units—wells and barriers—but subject to different boundary conditions. We explore these relations in this chapter. For the wells W and barriers B, we choose (V_j, δ_j) as follows:

$$
\begin{array}{lllll}
\text{Well}: & W & (9.0, 1.0) & (0.0, 8.0) & (9.0, 1.0), \\
\text{Barrier}: & B & (0.0, 4.0) & (9.0, 2.0) & (0.0, 4.0).
\end{array}
\tag{41.1}
$$

We first consider how energy eigenvalues interleave in binding potentials, scattering potentials, and periodic potentials. We then consider how these energies are related in similar binding, scattering, and periodic potentials.

41.1 INTERLEAVING OF BOUND STATE EIGENVALUES

The bound states of a single well W with asymptotic potentials $V_L = V_R = 9.0$ eV occur at ~ 1.8, 3.8, and 6.6 eV. In the triple well binding potential WWW with $V_L = V_R = 9.0$ eV each eigenvalue is split into a triplet. These triplets are shown in Fig. 41.1, on the left, over "Binding." The triplets extend over the two left wells shown in the three well binding potential.

If a fourth well is added (not shown), each singlet is split into a quartet of levels. The four levels obtained for the four well binding potential are shown extending over the two wells on the right of the three well potential. By comparing the triplet with the quartet, we observe:

- Between each adjacent pair of levels of the triplet, there is a single level of the quartet.

- Between each adjacent pair of levels of the quartet, there is a single level of the triplet.

- The separation between the two outer levels of the quartet is larger than the separation between the two outer levels of the corresponding triplet.

These observations are independent of the energy eigenvalue of the original single well, the shape of the individual well, and the number of wells. The N levels of an N-tuplet interleave the $N + 1$ levels of an $(N + 1)$-tuplet, for all N. The separation between the upper and lower levels of an N-tuplet increases as N increases, but this width is bounded above as $N \to \infty$. This means that the bound state levels of an N-tuplet become increasingly crowded as N increases. The upper bound on the width of an N-tuplet, as $N \to \infty$, is the width of the corresponding band of the corresponding periodic potential.

41.2   INTERLEAVING OF RESONANCE PEAKS

The resonance peaks of the double barrier B^2 with asymptotic potentials $V_L = V_R = 0.0$ eV occur approximately at the energies of the corresponding single well potential W: ~ 1.8, 3.8, and 6.6 eV. In the four barrier scattering potential $BBBB$ with $V_L = V_R = 0.0$ eV each resonance peak is split into a triplet. The energies at which these triplet peaks occur are shown in Fig. 41.1, in the middle, over "Scattering." The triplets extend over the two left wells shown in this scattering potential.

If a fifth barrier is added (not shown), each resonance peak is split into a quartet of resonances. The four resonances obtained for the five barrier scattering potential are shown extending over the two wells on the right. By comparing the resonance triplet with the resonance quartet, we observe:

- Between each adjacent pair of resonances of the triplet, there is a single resonance of the quartet.

- Between each adjacent pair of resonances of the quartet, there is a single resonance of the triplet.

- The separation between the two outer resonances of the quartet is larger than the separation between the two outer resonances of the corresponding triplet.

These observations are independent of the resonance of the original double barrier potential, the shape of the barriers, and the number of barriers. The N resonances of an N-tuplet interleave the $N + 1$ resonances of an $(N + 1)$-tuplet, for all N. The separation between the upper and lower resonances of an N-tuplet increases as N increases, but this separation is bounded above as $N \to \infty$. This means that the resonances of an N-tuplet become increasingly crowded as N increases. The upper bound on the width of an N-tuplet, as $N \to \infty$, is the width of the corresponding band of the corresponding periodic potential.

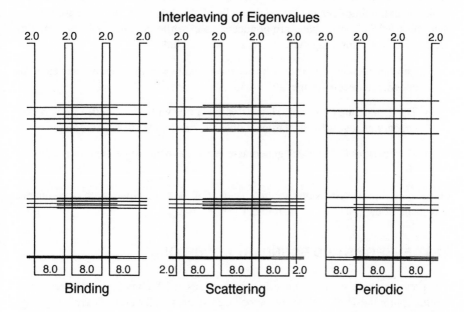

Fig. 41.1 *Binding*, The bound states in the single well split into into triplets of bound levels in the three well potential shown. The triplets are drawn over the two wells on the left. In the four well potential each bound state becomes a quartet. The quartet levels are drawn over the two wells on the right. *Scattering*, Each isolated resonance in a scattering potential containing a single well splits into a triplet or a quartet when the scattering potential contains three or four wells (four or five barriers). *Periodic*, In a periodic potential containing three wells, each band contains three levels. The lowest (highest) is nondegenerate in even (odd) bands. In the periodic potential W^4 the levels at the band edges are nondegenerate and the interior levels are doubly degenerate. The interior levels interleave, and the nondegenerate levels at the band edges occur at energies independent of N.

41.3   INTERLEAVING OF EIGENVALUES OF A PERIODIC POTENTIAL

It is possible to associate a periodic lattice with each of the binding potentials and scattering potentials just discussed. The binding potential W^N ($V_L = V_R = 9.0$ eV) and scattering potential B^{N+1} ($V_L = V_R = 0.0$ eV) are both associated with the periodic potential $W^N = B^N$. A periodic potential with $N = 3$ is shown in Fig. 41.1.

As discussed in chapter 38, the structure of the allowed energy eigenvalue spectrum differs slightly, depending on whether N is even or odd. The allowed eigenvalues for the periodic potential with $N = 3$ are shown extending over the left two wells shown in the three well periodic potential, shown in Fig. 41.1 on the right over "Periodic." For the even bands $(0, 2, \ldots)$ the lowest level is nondegenerate, while for the odd bands

$(1, 3, \dots)$ the upper level is nondegenerate. The eigenvalue spectrum for $N = 4$ is shown extending over the two right wells of this potential. In this case the two outermost levels in each band are nondegenerate, while the interior level is doubly degenerate. By comparing these two cases, we conclude:

- The degenerate ("interior") levels for the lattice with N cells interleave the degenerate levels for the lattice with $N + 1$ cells.

- The degenerate ("interior") levels for the lattice with $N + 1$ cells interleave the degenerate levels for the lattice with $N + 2$ cells, for all N.

- The energies of the nondegenerate levels at the band edges are independent of N.

Once again, these conclusions are independent of the parent eigenstate or resonance and shape of the potential in the unit cell.

41.4 ENERGIES AND BOUNDARY CONDITIONS

The previous three sections show that the energies at which bound states occur in a binding potential W^N with N wells behave systematically ("interleave") as N increases. Similar results apply to the transmission resonances for scattering potentials B^{N+1} and eigenvalues for periodic potentials $W^N = B^N$.

However, the energies at which bound states, resonance peaks, and eigenvalues of a periodic lattice occur are also closely related.

In Fig. 41.2 we compare the energy eigenvalues for the two potentials W^3 and W^4 ($V_L = V_R = 9.0$ eV) and the two scattering potentials B^4 and B^5 ($V_L = V_R = 0.0$ eV). In this figure we plot the bound state triplet and quartet (over Binding). These two multiplets are offset horizontally so they can be distinguished. These levels interleave. We have also plotted the energies at which the triplet and quartet of resonance peaks occur (over Scattering). These two multiplets are also slightly offset. These energies also interleave.

The multiplets have been offset in such a way that the quartet of the binding (W^4) and scattering (B^5) potentials slightly overlap. It is clear that each of the four levels in the binding potential (W^4) lies slightly lower in energy than the corresponding resonance in the scattering potential (B^5). The same is true for the levels in the bound state triplet arising from W^3 and the resonance triplet arising from B^4.

As N increases, the bound state levels and scattering resonances are squeezed closer together. The lowest bound state level approaches the bottom of the energy band of the periodic lattice, and the highest of the scattering resonance peaks approaches the top of the energy band. The band edges for the periodic potential are shown in Fig. 41.2 (over Periodic).

These numerical results were computed for the particular well and barrier configurations described above (in equation (41.1)), and for the third bound state (resonance, energy band) at ~ 6.6 eV. However, the results are independent of the shape of the well or barrier and also independent of the original parent level chosen for this comparison.

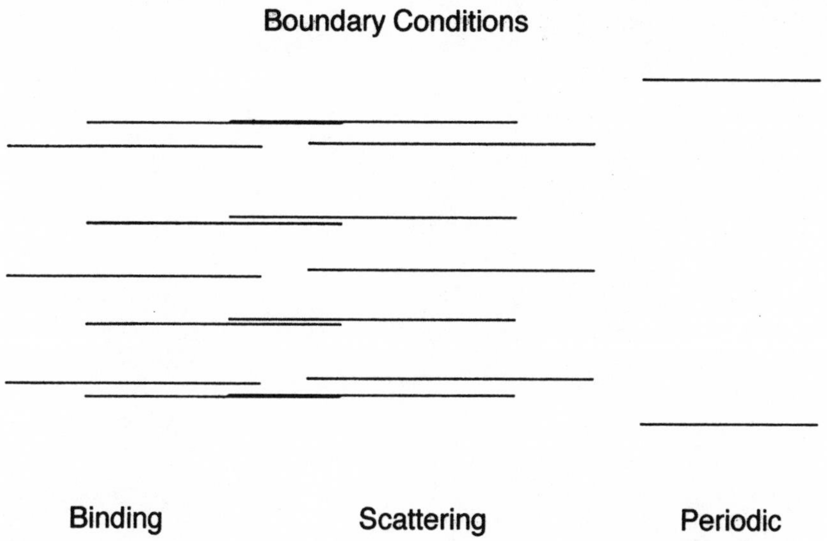

Boundary Conditions

Binding   Scattering   Periodic

Fig. 41.2 The triplet and quartet of bound state energies (over Binding) are offset horizon-tally. They interleave. Similarly for the resonance peaks (over Scattering). Each resonance peak lies slightly above the corresponding bound state, both for the quartet and the triplet. All levels and resonances lie within the edges of the corresponding band (shown over Periodic) of the corresponding periodic potential.

42

Wavefunctions and Probability Distributions

The transfer matrices $T = M_1 M_2$, which were used in chapters 37 and 38, relate the amplitudes of the trigonometric/hyperbolic cosines and sines in adjacent unit cells of a periodic lattice. We make this point explicit here. Suppose unit cell i consists of two regions, 1 and 2, in which the potential is constant, as shown in Fig. 42.1. We choose $V_1 = -5.0$ eV, $V_2 = 0.0$ eV and $V_1 < E < V_2$. In regions 1 and 2 of cell i and region 1 of cell $i + 1$ the wavefunctions are

$$
\begin{aligned}
\psi_i^1(x) &= A_i \cos kx &+& \quad B_i \tfrac{1}{k} \sin kx \,, \\
\psi_i^2(x) &= C_i \cosh \kappa x &+& \quad D_i \tfrac{1}{\kappa} \sinh \kappa x \,, \\
\psi_{i+1}^1(x) &= A_{i+1} \cos kx &+& \quad B_{i+1} \tfrac{1}{k} \sin kx \,.
\end{aligned}
\tag{42.1}
$$

Within each region, the value of x ranges from 0 at the left edge to $L_1 = 8.0$ Å in region 1 and $L_2 = 2.0$ Å in region 2. When the boundary conditions are imposed, we obtain the following equations:

$$
\begin{pmatrix} \cos kL_1 & \tfrac{1}{k} \sin kL_1 \\ -k \sin kL_1 & \cos kL_1 \end{pmatrix}
\begin{pmatrix} A \\ B \end{pmatrix}_i
= \begin{pmatrix} C \\ D \end{pmatrix}_i ,
\tag{42.2}
$$

$$
\begin{pmatrix} \cosh \kappa L_2 & \tfrac{1}{\kappa} \sinh \kappa L_2 \\ \kappa \sinh \kappa L_2 & \cosh \kappa L_2 \end{pmatrix}
\begin{pmatrix} C \\ D \end{pmatrix}_i
= \begin{pmatrix} A \\ B \end{pmatrix}_{i+1} .
\tag{42.3}
$$

These equations are easily inverted

$$
\begin{pmatrix} A \\ B \end{pmatrix}_i
= \begin{pmatrix} \cos kL_1 & -\tfrac{1}{k} \sin kL_1 \\ k \sin kL_1 & \cos kL_1 \end{pmatrix}
\begin{pmatrix} C \\ D \end{pmatrix}_i
= M_1 \begin{pmatrix} C \\ D \end{pmatrix}_i ,
\tag{42.4}
$$

$$\begin{pmatrix} C \\ D \end{pmatrix}_i = \begin{pmatrix} \cosh \kappa L_2 & -\frac{1}{\kappa}\sinh \kappa L_2 \\ -\kappa \sinh \kappa L_2 & \cosh \kappa L_2 \end{pmatrix} \begin{pmatrix} A \\ B \end{pmatrix}_{i+1} = M_2 \begin{pmatrix} A \\ B \end{pmatrix}_{i+1}.$$
(42.5)

As a result

$$\begin{pmatrix} A \\ B \end{pmatrix}_i = M_1 M_2 \begin{pmatrix} A \\ B \end{pmatrix}_{i+1} = T \begin{pmatrix} A \\ B \end{pmatrix}_{i+1}.$$
(42.6)

Wavefunctions in a Lattice

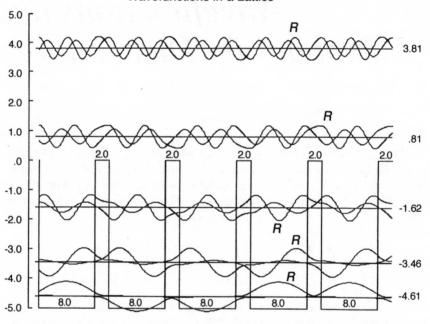

Fig. 42.1 Real and imaginary parts for one eigenstate in each of the five lowest bands of the periodic potential shown. The real parts are identified.

Remark. If we had used (A, B) as amplitudes for the right- and left-propagating exponentials e^{+ikx}, e^{-ikx} and (C, D) as amplitudes for the exponentially decaying and growing exponentials $e^{-\kappa x}, e^{+\kappa x}$, then we would have found the following relation

$$\begin{pmatrix} A \\ B \end{pmatrix}_i = K^{-1}TK \begin{pmatrix} A \\ B \end{pmatrix}_{i+1}, \qquad K = \begin{bmatrix} 1 & 1 \\ +ik & -ik \end{bmatrix}.$$
(42.7)

The quantization condition (i.e., (37.3)) is valid whether we use T or its similarity transformed version $K^{-1}TK$. However, wavefunction calculations are *much* easier with the cosine and sine basis set.

42.1   UNIT CELL PHASE SHIFT

The wavefunctions in any part of cell i and the corresponding part of cell $i+1$ are

$$\psi_i(x) \;=\; \big(\; \Phi_1(x) \;\; \Phi_2(x) \;\big) \begin{pmatrix} A \\ B \end{pmatrix}_i,$$

$$\psi_{i+1}(x) \;=\; \big(\; \Phi_1(x) \;\; \Phi_2(x) \;\big) \begin{pmatrix} A \\ B \end{pmatrix}_{i+1} \tag{42.8}$$

$$\;=\; \big(\; \Phi_1(x) \;\; \Phi_2(x) \;\big) T^{-1} \begin{pmatrix} A \\ B \end{pmatrix}_i.$$

The last equation was obtained by inverting (42.6). If the amplitudes (A,B) are chosen as eigenvectors of T with eigenvalue λ, then $T^{-1}\begin{pmatrix} A \\ B \end{pmatrix} = \lambda^{-1}\begin{pmatrix} A \\ B \end{pmatrix}$. As a result

$$\psi_{i+1}(x) = \lambda^{-1}\psi_i(x). \tag{42.9}$$

In an allowed band, $\lambda = e^{\pm i\phi}$. This means that the wavefunction in cell $i+1$ is simply a phase-shifted version of the wavefunction in cell i. If the wavefunction in cell i is known, it can be used to construct the wavefunction in the adjacent cells. For example, if we write the wavefunctions in terms of their real and imaginary parts and choose $\lambda^{-1} = e^{i\phi}$, then

$$\psi^R_{i+1}(x) + i\psi^I_{i+1}(x) = e^{i\phi}\left(\psi^R_i(x) + i\psi^I_i(x)\right), \tag{42.10}$$

$$\begin{pmatrix} \psi^R(x) \\ \psi^I(x) \end{pmatrix}_{i+1} = \begin{pmatrix} \cos\phi & -\sin\phi \\ \sin\phi & \cos\phi \end{pmatrix} \begin{pmatrix} \psi^R(x) \\ \psi^I(x) \end{pmatrix}_i. \tag{42.11}$$

42.2   WAVEFUNCTIONS

The wavefunction in any part of cell i can be written as a linear superposition of the basis functions $\Phi_1(x), \Phi_2(x)$ in that region:

$$\psi_i(x) = A_i\Phi_1(x) + B_i\Phi_2(x), \tag{42.12}$$

where the amplitudes $(A,B)_i$ satisfy the appropriate eigenvalue equation:

$$T\begin{pmatrix} A \\ B \end{pmatrix}_i = e^{i\phi}\begin{pmatrix} A \\ B \end{pmatrix}_i.$$

To compute the unnormalized amplitudes $(A, B)_i$, we rewrite this equation as follows

$$\begin{bmatrix} t_{11} - e^{i\phi} & t_{12} \\ t_{21} & t_{22} - e^{i\phi} \end{bmatrix} \begin{bmatrix} A \\ B \end{bmatrix} = 0 \iff \begin{array}{rcl} (t_{11} - e^{i\phi})A + t_{12}B & = & 0 \\ t_{21}A + (t_{22} - e^{i\phi})B & = & 0 \end{array} .$$

$$(42.13)$$

The two equations on the right are not independent. They can be used to determine A and B

$$\begin{pmatrix} A \\ B \end{pmatrix} \sim \begin{pmatrix} e^{i\phi} - t_{22} \\ t_{21} \end{pmatrix} \sim \begin{pmatrix} t_{12} \\ e^{i\phi} - t_{11} \end{pmatrix} . \qquad (42.14)$$

The two expressions for the amplitudes are proportional to each other. Using the first expression, we find

$$\psi_i(x) = \underbrace{\frac{1}{2}(t_{11} - t_{22}) \cos kx + t_{21}\frac{1}{k} \sin kx}_{\psi_i^R(x)} + i\underbrace{\sin \phi \, \cos \, kx}_{\psi_i^I(x)} . \qquad (42.15)$$

The wavefunctions in other regions of the ith cell are computed using the transfer matrices M_1, M_2, and so on (see equation (42.6)). The wavefunctions in adjacent regions are computed using the phase shift property (42.10).

In Fig. 42.1 we show wavefunctions computed for one state in each of the five lowest bands in the periodic potential shown. The real and imaginary parts of each wavefunction are plotted and are labeled with an R or an I. All wavefunctions have been normalized so that the integral of $|\psi(x)|^2$ over a unit cell is the same for each eigenfunction.

In fact, the real and imaginary parts of these wavefunctions are interchangeable. If a wavefunction obtained using the angle $-\phi$ is multiplied by i, then we obtain

$$i \times \left(\psi^R(x) - i\psi^I(x)\right) = \psi^I + i\psi^R(x) . \qquad (42.16)$$

This is the wavefunction obtained using the angle ϕ and interchanging the real and imaginary parts.

The nondegenerate eigenfunctions at the edges of each band can always be made real, since the eigenvalues are $\lambda = \pm 1$. The corresponding wavefunctions are standing waves, whereas the degenerate eigenfunctions describe waves traveling to the left and right.

42.3   PROBABILITY DISTRIBUTIONS

Since the wavefunction in cell $i+1$ is just a phase-shifted version of the wavefunction in cell i, the probability distribution in cell $i+1$ is equal to the probability distribution in cell i:

$$P_{i+1}(x) = |\psi_{i+1}(x)|^2 = |e^{i\phi}\psi_i(x)|^2 = |e^{i\phi}|^2 \, |\psi_i(x)|^2 = |\psi_i(x)|^2 = P_i(x) \,.$$
(42.17)

As a result, the probability distribution is invariant from cell to cell.

In Fig. 42.2 we plot the probability distributions for one eigenstate in each of the five lowest bands of the potential shown in Figs. 42.1 and 42.2. These distributions have all been properly normalized (the integral over every unit cell is $1/N$ for each of the probabilities). Although the nodal structure of the wavefunctions is not obvious from Fig. 42.1, it is much clearer in Fig. 42.2. The number of "nodes" within each well increases by one for each successive band. The energies at which the computations were done for Figs. 42.1 and 42.2 differ slightly.

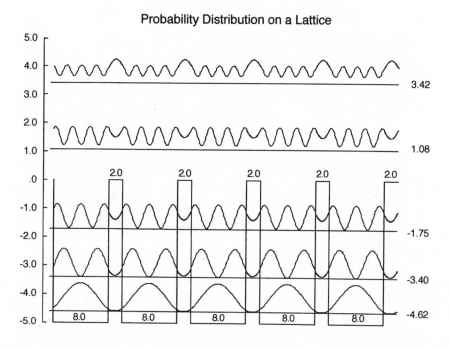

Fig. 42.2 Probability distributions for one eigenstate in each of the five lowest bands of the potential shown. The number of "nodes" increases with band number. It is clear that in an eigenstate, the probability distribution is the same in each cell of the lattice.

Remark: For $E < 0$ the bound state wavefunctions in Fig. 42.2 are real and exhibit nodes in the wells. The probability distribution goes to zero at the nodes. The lowest three states have zero-, one-, and two nodes. For $E > 0$ the wavefunctions do not have nodes (they are complex). The lowest two states with $E > 0$ have three- and four "pseudo-nodes," where the probability distribution has deep, nonzero minima in the wells.

43

Alloys

An alloy consists of a homogeneous mixture of two or more substances, usually metals. Energy bands can be computed for alloys just as they can be computed for pure substances.

In this chapter we will consider energy bands only for the simplest one-dimensional alloys. The alloys we consider are composed of equal numbers of two substances A and B in a periodic lattice of type $(AB)^N$. In Fig. 43.1 (top left) we show the potential for the unit cell A, with $(V, \delta) = (-5.0, 8.0), (0.0, 2.0)$. The bound states for the corresponding single well potential are shown. They occur at $\sim -4.6, -3.4$, and -1.6 eV. To the right of this potential we show the density of states for the lattice A^N. The three bound states have been spread out into three bands. The density of states also shows a fourth band from ~ 0.2 to ~ 1.7 eV that arises from the first transmission resonance, and the lower part of a fifth band that arises from the second transmission resonance of the single well potential.

The potential for the unit cell B $((V, \delta) = (-2.0, 3.0), (0.0, 1.0))$ is shown on the second line of Fig. 43.1. The single bound state for the corresponding binding potential occurs just above -1.0 eV. To the right of this potential is the density of states plot for the periodic lattice B^N. The band arising from the single bound state extends from -1.6 to $+0.4$ eV. The upper levels in this band are unbound, even though the band arises from a bound state. The second band, extending above 1.2 eV, arises from the lowest transmission resonance.

The potential for the unit cell AB is shown on the bottom line of Fig. 43.1. This potential has four bound states. The two lower ones are essentially where they were in the pure A binding potential. The two upper levels have repelled each other slightly from their original positions in the A well and the B well. The third bound state is

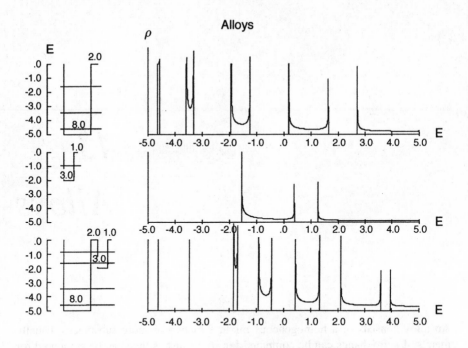

Fig. 43.1 *Top*, Potential A, eigenstates of corresponding single well potential, and density of states for corresponding periodic lattice A^N. *Middle*, Similar to top line, except for potential B. *Bottom*, Potentials A and B form the unit lattice potential AB. Eigenvalues and density of states for corresponding potential are shown.

mostly confined to the A well and the fourth is mostly confined to the B well. This was determined by computing the eigenfunctions for the AB well.

The density of states for the periodic lattice $(AB)^N$ is shown to the right of the AB potential. The two lowest bands, derived from the A well, have been squeezed and are now so narrow that they cannot be resolved. These two bands have been squeezed because the classically forbidden region between adjacent A wells has been greatly increased by inserting the B potential. The third A band (-1.95 to -1.3 eV) has been squeezed to the region of ~ -1.8 eV, while the lowest B band (-1.6 to $+0.4$ eV) has been squeezed into (-0.9 to -0.5 eV), below the ionization threshold. The three bands that appear above 0.0 eV are not confined primarily to either the A well or the B well.

These results, and the conclusions drawn from them, are not strongly dependent on the actual shapes of the potentials. In Fig. 43.2 we repeat the calculations shown in Fig. 43.1 for smoother versions of potentials A and B. The potentials A and B of Fig. 43.1 have been replaced by smoother potentials with essentially the same bound state spectra. The major differences are that the third band of the modified A potential extends above the ionization threshold while the first band of the modified B potential does not. The two lowest bands of the modified $(AB)^N$ potential are not as narrow as the corresponding bands in the original $(AB)^N$ potential (Fig. 43.1) because the classically forbidden region between two adjacent A wells is smaller. Otherwise, the density of states for the periodic potentials $(AB)^N$ in Fig. 43.1 and Fig. 43.2 are very similar.

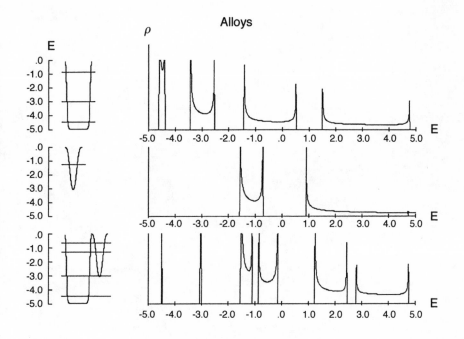

Fig. 43.2 Similar to Fig. 43.1 but with smoothed versions of potentials A and B, as shown. The modified potentials have bound state spectra similar to the unmodified potentials.

44

Superlattices

A one-to-one alloy containing unit cells of type A and B in a regular lattice has the one-dimensional spatial structure

$$\ldots A\,B\,A\,B\,A\,B\,A\,B\,A\,B\,A\,B\,A\,B\,A\,B\,A\,B\,A\,B \ldots . \qquad (44.1)$$

Imperfections in this regular structure often occur. They may take the form of inserting or deleting atoms of type A or B at various points in the lattice. In this chapter we consider imperfections that are caused by deleting atoms. The deletions involve alternating atoms of type A and B spaced at regular intervals. These deletions lead to regular lattices with a more complicated structure than the original lattice. For example, if every third atom is deleted from the regular lattice with unit cell (AB),

$$\ldots A\,B\,\underline{A}\,B\,A\,\underline{B}\,A\,B\,\underline{A}\,B\,A\,\underline{B}\,A\,B\,\underline{A}\,B\,A\,\underline{B}\,A\,B \ldots = [(AB)(BA)]^N ,$$
$$(44.2)$$

we find a regular lattice with unit cell $[(AB)(BA)]$. A lattice of this type is called a *superlattice*.

Other superlattices can be obtained by deleting every fifth, seventh, $\ldots$, $(2d+1)$th atoms

d	$2d+1$	Regular Lattice
1	3	$[(AB)^1(BA)^1]^{N/4}$
2	5	$[(AB)^2(BA)^2]^{N/8}$
3	7	$[(AB)^3(BA)^3]^{N/12}$
d	$2d+1$	$[(AB)^d(BA)^d]^{N/4d}$.

$$(44.3)$$

Each of the superlattices above has a total of N atoms, half of type A and half of type B. Since a superlattice obeys periodic boundary conditions, the allowed energy levels occur in bands. It may be expected that the band structure of a superlattice is closely related to the band structure of the parent lattice $(AB)^{N/2}$.

To explore this question, we can compute the band structure of a series of super-lattices with increasing d. We choose unit cell parameters as follows

	A		B	
	V	δ	V	δ
Region 1	0.0	1.0	0.0	0.5
Region 2	−5.0	8.0	−6.5	3.0
Region 3	0.0	1.0	0.0	0.5

$$(44.4)$$

In Fig. 44.1 we describe the changes that take place in a single band of the original $(AB)^{N/2}$ lattice when the alloy is converted to a superlattice with the same total number of atoms, as a function of increasing d. The $(AB)^{N/2}$ band that is presented in Fig. 44.1 extends from 3.5 to 4.6 eV. As usual, all bands behave in the same way, and this behavior is independent of the details of the potentials of A and B.[1]

The regular lattice $(AB)^{N/2}$ has a band extending from 3.5 to 4.6 eV. For the superlattice with $d = 1$, the original band is split into two subbands. The two subbands are separated by a rather large band gap.

For the superlattice with $d = 2$, the original band has separated into four subbands. The two outer subbands are separated from the inner two by large band gaps. The inner two are separated by a very small gap at 4.1 eV.

In the superlattices with $d = 3, 4$, and 5, the original band has split into six, eight, and ten subbands. For $d = 3$ the six subbands are separated by three large and two small gaps at 3.8 and 4.45 eV. For $d = 4$ the eight subbands are separated by four large and three small gaps. For $d = 5$ there are five large and four small gaps.

In the case of the superlattice $[(AB)^d(BA)^d]^{N/4d}$, the original band has split into $2d$ subbands that are separated by $2d - 1$ band gaps, of which d are large and $d - 1$ are small. The large and small band gaps alternate with each other.

The structure of superlattice subbands can be viewed from a slightly different perspective. In going from the regular lattice $(AB)^{N/2}$ to the $d = 1$ superlattice $[(AB)^1(BA)^1]^{N/4}$ the original band splits into two subbands. In going from the $d = 1$ to the $d = 2$ superlattice, a new subband appears inserted between the two $d = 1$ subbands, squeezing these two outer subbands further past the band edges of the original $(AB)^{N/2}$ band. This new subband, extending from 3.8 to 4.5 eV, actually consists of two subbands separated by a narrow gap at $\sim$4.1 eV.

In going from the $d = 2$ to the $d = 3$ superlattice, the narrow gap at $\sim$4.1 between the two middle $d = 2$ subbands grows wider, extending from $\sim$4.0 to 4.2 eV. Each of the two new well-separated subbands (3.7 to 4.0 eV and 4.2 to 4.7 eV) is split by a narrow gap (at 3.8 and 4.45 eV, respectively).

[1]S. M. Bose, D. H. Feng, R. Gilmore, and S. Prutzer, "Electronic density of states of a one-dimensional superlattice," *Journal of Physics F* 10 (1980): 1129–1133.

In progressing from the $d = 3$ to the $d = 4$ superlattice, a new band is inserted in the middle of the subband structure of the $d = 3$ superlattice. This new band extends from ~ 3.9 to 4.3 eV and actually consists of two subbands split by a narrow gap at ~ 4.1 eV.

The changes in the subband structure in going from the $d = 4$ to the $d = 5$ superlattice are similar to the changes that take place in going from the $d = 2$ to the $d = 3$ superlattice.

As Fig. 44.1 makes clear, it is useful to regard each band of the $(AB)^{N/2}$ lattice as breaking up into $d + 1$ "fat subbands" in the $[(AB)^d(BA)^d]^{N/4d}$ superlattice. The two outer fat subbands are not split while the inner $d - 1$ "fat subbands" each consist of two subbands separated by a narrow gap.

When the number of fat subbands is even ($d = 1, 3, 5, \ldots$), a new fat subband is inserted right in the middle of the fat subbands in going from the d to the $d + 1$ superlattice. When the number of fat subbands is odd ($d = 2, 4, \ldots$), then in going from the d to the $d + 1$ superlattice, the middle fat subband of the d superlattice "splits." That is, the narrow band gap in that fat subband become large, and each of the two resulting subbands develops a narrow band gap.

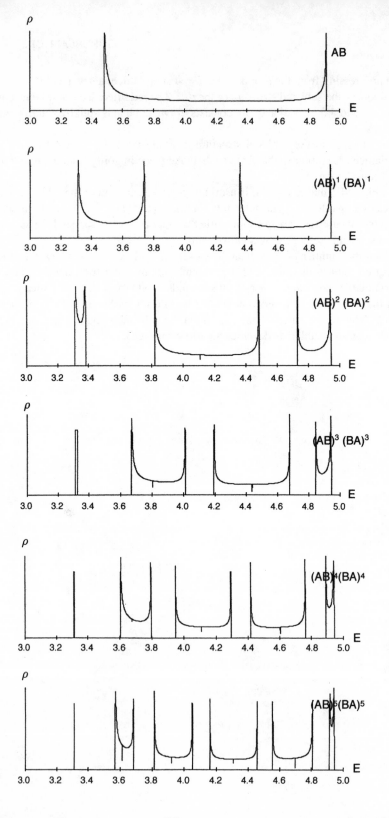

Fig. 44.1 Band structure of the $(AB)^{N/2}$ lattice, and the superlattices $[(AB)^d(BA)^d]^{N/4d}$ for $d = 1, 2, 3, 4, 5$.

45

Impurities

In the previous chapters we have seen how energy levels in periodic potentials of the type A^N occur in bands, with each band containing a total of N states. The same is true for regular alloys of type $(AB)^{N/2}$, as well as superlattices of type $[(AB)^d(BA)^d]^{N/(4d)}$.

It is difficult to grow crystal lattices of very high purity. Usually a crystal lattice contains impurities. We should therefore ask:

- Does the energy band structure of a pure lattice A^N persist in a lattice with impurities: $A^{N-1}B$?

- How are the electron wavefunctions affected by the presence of impurities?

These questions can be addressed relatively easily if we use what we have learned in previous chapters to simplify our computations as much as possible.

We exploit the following observations

- The band structure of a regular lattice A^N (N large) is apparent even for small values of N ($N \sim 5$).

- The band structure is unaffected by the type of boundary conditions imposed (bound state, scattering, periodic), for all practical purposes.

These observations suggest that we can determine the effects of impurities on the band structure and wavefunctions of a regular lattice by computing the band structure and wavefunctions for the bound states of a relatively small system $A^{N-1}B$, (N small).

To explore the effect of an impurity atom B on the properties of a regular lattice A^N, we have computed the bound state spectrum for two lattices of type A^3BA. The

207

results are shown in Fig. 45.1 and Fig. 45.2. In both figures the potential B depends on a parameter.

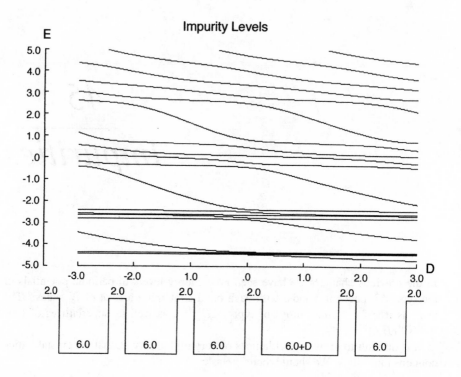

Fig. 45.1 Spectrum of bound state levels for the potential A^3BA as the width of the B impurity is varied. The A bands are represented by quartets of levels. Note that the eigenvalues do not cross. For this potential $V_L = V_R = 10.0$ eV.

In Fig. 45.1 both potentials A and B are 5.0 eV deep. The tops of the potentials are at 0.0 eV and the bottoms are at -5.0 eV. The width of potential A is 6.0 Å while B has a width of 6.0+D Å. The barriers separating wells are uniformly thick, at 2.0 Å. The asymptotic potentials V_L and V_R are 10.0 eV to ensure that we see only bound states in the range scanned (-5.0 to $+5.0$ eV). For $D = 0$ the lattice is A^5, and the bound state energies below 5.0 eV are gathered into four groups of quintets. We can easily see the effect that varying the width of B has on the energy-level structure. For D negative, the narrow B well has energies higher than the energies in a single A well. As D is increased, the lowest B level settles into the top of the lowest A band. The next B level, pressed up against the bottom of the third A band, peels off and settles onto the top of the second A band. The next higher B level behaves similarly. For $D > 0$, the B well is wider than the A well and its spectrum has lower-lying levels. This behavior is reflected in Fig. 45.1. As D increases through

0.0 Å, the lowest-lying level in each band falls away from the bottom of the band and approaches the top of the next lower A band. Of course, the lowest B level has no lower A band to settle onto.

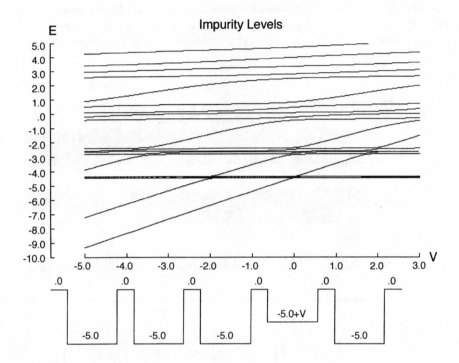

Fig. 45.2 Spectrum of bound state levels for the potential A^3BA as the depth of the B impurity is varied. The A bands are represented by quartets of levels. Note the the eigenvalues do not cross. For this potential $V_L = V_R = 10.0$ eV.

Fig. 45.2 tells a similar story. Here the wells all have the same width (6.0 Å). The barriers at 0.0 eV have width 2.0 Å. Each A well is 5.0 eV deep, while the bottom of the B well is at $-5.0 + V$ eV. The B well is varied in depth from -10.0 eV ($V = -5.0$ eV) to -2.0 eV ($V = +3.0$ eV). Once again, the band structure of A^N is represented by four groups of quintets for A^5 at $V = 0.0$ eV. As the bottom of the B well is increased from -10.0 eV to -2.0 eV, the bound state energies due to B increase almost linearly (with a slope of $+1$). However, the B levels do not cross the A bands. Rather, a B level approaches the bottom of an A band, and then, at a slightly greater value of V, the highest level in the A band peels off and rises linearly with increasing V until it runs into the bottom of the next higher A band.

These two figures make it clear that, in some sense, there is a "conservation of number of levels" in going from a regular lattice A^N to a lattice with impurity $A^{N-1}B$ (or some cyclic permutation). The band structure is not lost, merely perturbed. Bands with N levels in A^N are replaced by bands with $N-1$ levels in $A^{N-1}B$, and additional B levels are scattered in appropriate places in the energy-level spectrum.

The energy eigenfunctions are perturbed as follows. For eigenvalues in a band of levels, the corresponding wavefunctions extend over all the A atoms. For any eigenstate in an A band, the probability distribution is essentially the same in every cell of type A, except possibly in the A cells adjacent to the B impurity. However, the nondegenerate states between A bands have wavefunctions that are largely localized to the B atom impurity.

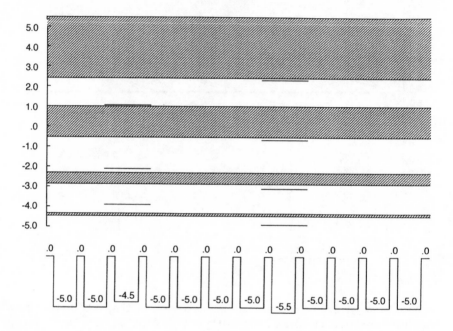

Fig. 45.3 Bands and impurity levels in a lattice of type $A^{n_1}BA^{n_2}B'A^{n_3}$. The bands are shown extending over the entire lattice. The donor level, over B' with depth 5.5 eV, lies slightly below the lowest unfilled or partly filled A band. The acceptor level, over B with depth 4.5 eV, lies slightly above the highest filled or almost filled A band.

In Fig. 45.3 we show a lattice $A^{n_1}BA^{n_2}B'A^{n_3}$. The A levels occur in the bands indicated. Since the A-like eigenstates extend over the entire lattice, the bands are shown extending over the lattice. The B level at -4.5 eV provides three levels that occur above the A bands. Since the wavefunctions, with $n = 0, 1, 2$ nodes, are largely confined to this well, the energy level is shown extending over only this well and its adjacent barriers. The B' level at -5.5 eV provides four levels that occur beneath the A bands. Since the wavefunctions, with $n = 0, 1, 2, 3$ nodes, are largely confined to this well, the energy level is shown extending over this B' well and its adjacent barriers.

Impurity levels that occur below an energy band that contains few or no electrons are called donor levels. If these impurity levels contain an electron, that electron is "stuck" to the impurity site. However, that electron can be promoted (donated) into the empty band just above the donor level at relatively low energy cost. In the empty band, the electron is mobile: it can move relatively freely from A atom to A atom. The level of the B' well (with depth 5.5 eV) just below the lowest empty A band is a donor level.

An empty impurity level just above a filled A band is called an acceptor level. At relatively low energy cost, this level can remove (accept) an electron from the filled band just below it. This leaves a "hole" in the filled band. This hole acts in many ways like an electron of opposite charge. The level of the B well (with depth 4.5 eV) just above the highest-filled A band is an acceptor level.

Careful control of the concentrations and types (donor, acceptor) of dopants in semiconductors have been responsible in large part for the current computer revolution.

46

Quantum Engineering

The availability of energy resources will be one of the major points of political friction in the twenty-first century.

46.1 ENERGY SOURCES

The energy resources that currently contribute more than an infinitesimal fraction of the world's energy needs are listed in Table 46.1.

Coal, oil, and gas are the major sources of energy in today's economy. They are "nonrenewable" and are currently being rapidly depleted. We believe these energy sources are of fossil origin. This means that the energy stored in these resources derives ultimately from the transformation of solar energy to biomass, at very low efficiency, followed by death, sedimentation, conversion, and storage (also at very low efficiency) over a very long period of time. These resources have been created by natural processes that continue to this day. The time scale for creation of these resources is millions of years; the time scale for depletion is hundreds of years. Therefore, although these resources are in principle renewable, they are in practice nonrenewable unless we learn how to accelerate the creation process by factors in excess of 10^4. Burning these fossil fuels releases large amounts of CO_2 and other pollutants into the atmosphere.[1] It is widely believed that this has a negative environmental impact. In addition, it destroys a reservoir of raw materials for industry.

---

[1]M. I. Hoffeert et al., "Advanced Technology Paths to Global Climate Stability: Energy for a Greenhouse Planet," *Science* 298 (2002): 981-87.

Table 46.1   Energy resources

Energy Source	Comment
Coal	"Nonrenewable"
Oil	
Gas	
Fission	Radioactive
Fusion	
Geothermal	
Tidal	Gravitational
Hydroelectric	Gravitational/solar
Wind	Solar
Biomass	
Solar	

Some people regard this loss as even more severe than possible negative impact on our atmosphere.

The next three energy sources listed in Table 46.1 involve nuclear transformations. Nuclear fission ("splitting") involves splitting heavy nuclei into lighter nuclei, releasing energy. This process, initially demonstrated on the earth in atomic bomb explosions, has successfully been tamed. It is now the basis of energy production in nuclear power plants that generate a substantial fraction of electrical energy in the industrial nations of the earth. When not handled carefully, nuclear accidents can occur (Three Mile Island, Chernobyl, ...). The worst accidents have lead to environmental catastrophes. That is, large areas are rendered uninhabitable for long periods by high levels of radiation. Although the radioactivity levels gradually die away (along with the population), the half-life for decay is comparable to or longer than the human time scale ($\sim$70 years). This renders fission a problematic energy source for the future.

Nuclear fusion involves combining lighter elements to form heavier elements, releasing energy in the process. Fusion of hydrogen to helium,

$$
\begin{aligned}
p^+ + p^+ & \rightarrow & d^+ + e^+ + \nu_e \,, \\
d^+ + p^+ & \rightarrow & {}^3He^{++} + \gamma \,, \\
{}^3He^{++} + {}^3He^{++} & \rightarrow & {}^4He^{++} + 2e^+ + 2\nu_e \,,
\end{aligned}
\tag{46.1}
$$

is the principle fusion process that powers our star, the sun. This process was demonstrated on the earth with the explosion of the first hydrogen bomb. Since then, it has been a goal to tame this reaction and allow it to run in a controlled, continuous way in order to generate energy for the world's needs. An enormous amount of time, money, and intellectual energy has been devoted to this goal. Estimates of the amount of time and money required to reach a commercially viable result have diverged rather than converged—the end is nowhere in sight and receding further from reach all the time.

Geothermal energy is a natural resource that is derived in second order from nuclear fission. Long-lived radioactive isotopes within the core and mantle of the earth decay, releasing energy that heats up the surroundings. The hot plastic material of the core and mantle flows with a geological time scale. This material comes close to the surface at some parts of the earth's surface, principally at boundaries between the earth's plates. The temperature difference between the hot interior magma and the cooler fluids on the earth's surface can be used in the usual way to generate heat and electrical energy. Although this resource is nonrenewable, its time scale is so long that, for all practical purposes, it can be considered as permanent.

The motion of the earth and moon about their center of mass, and of the earth-moon system about the sun, raises diurnal tides on the surface of the earth. The differing physical properties of the material on the earth's surface (water, rock) allows us to create differences in gravitational potential energy that ultimately can be transformed to electrical energy. Tidal energy will never provide more than a small percentage of the earth's energy needs. It is essentially perpetual and nonpolluting. However, it does require modification of parts of the earth's surface and in that sense does have a nontrivial environmental impact.

Hydroelectric energy is derived from the conversion of the energy in falling water into electrical energy. Water is raised from sea level by evaporation (solar energy) and deposited at higher altitudes by precipitation as rain or snow. At higher elevations it possesses gravitational potential energy. Since this energy conversion process occurs naturally,

> All the rivers run into the sea;
> Yet the sea is not full;
> Unto the place from whence the rivers come,
> Thither they return again.   Ecclesiastes 1:7

Hydroelectric conversion is nonpolluting. It is also "perpetual," or at any rate will last as long as there is water on the surface of the earth. However, it also involves modifying selected parts of the earth's surface, so in this sense it also has nontrivial environmental impact.

Wind motion and biomass conversion are energy sources that originate almost entirely with the conversion of incident solar radiation. Since they are also naturally occurring processes, they are renewable and nonpolluting. However, they also involve modification of the environment, so to some extent are subject to political and environmental constraints.

Tidal, hydroelectric, and wind power sources will not provide more than a small percentage of the earth's energy requirements. Biomass conversion has the potential to make a more substantial contribution, although it is not likely to.

The sun is ultimately the source of all but the radioactive energy sources. Solar energy can be converted directly to industrially useful energy in two ways (neglecting conversion to biomass):

1. directly to heat

2. directly to electrical energy.

In the former case, a large number of concentrating mirrors, spread over an area the size of one or more soccer fields, are used to focus sunlight into a very small volume. Within this volume the temperature can reach 5000 K. Such solar furnaces can be used to drive electrical turbines or used as research facilities. More mundane but commercially much more important applications involve absorbing sunlight to produce hot (80°C) water rather than wasting electricity to heat water.

Sunlight can be transformed directly to electrical energy by being absorbed in crystals that are designed to allow the absorbed energy to separate positive from negative charges. Solar panels composed of such crystals are routinely used in the space program. They are also used, with increasing frequency as costs drop in places where cost or maintenance is more of a problem for other types of energy sources.

We will describe how to design crystals for direct solar conversion later in this chapter.

46.2 ENERGY CONSUMPTION

The total energy that has been consumed by the earth's population in the year 2000 has been estimated to be about 500 exajoules (500×10^{18} J). It is convenient to convert this energy consumption per year into power measured in Watts (joules/sec). This is done by dividing by the number of seconds in a year. This is 3.15×10^7 sec (mnemonic: $\pi \times 10^7$ sec). The rate of energy consumption in 2000 is then about 1.58×10^{13} W.

We compare this energy consumption rate with the rate at which solar radiation energy is incident at the top of the earth's atmosphere. This number, the "solar constant," is 1368 W/m^2 according to earth satellite measurements.[2] This number is not actually constant: it varies primarily because of small variations in solar energy output (fluctuations of order 0.01%, or one part in 10^4, are observed) and variation in the earth-sun distance due to the eccentricity of the earth's orbit. For this reason, we will refer to this measured value as the "solar irradiance." The total power received by the earth at its mean solar distance is

$$P = 1368 \text{W/m}^2 \times \pi (6370 \text{km})^2 = 1.74 \times 10^{17} \text{W} , \qquad (46.2)$$

where the earth's radius is approximately 6370 km. About $\frac{3}{4}$ of this energy filters down to the surface of the earth as radiation in the visible part of the spectrum.

It is instructive to compare the rate at which energy is consumed by the world's population to the rate at which radiant energy is received from the sun by the earth. This ratio is

[2]K. J. H. Phillips, *Guide to the Sun*, Cambridge, UK: Cambridge University Press, 1992.

$$R = \frac{\text{Power consumed}}{\text{Power received}} = \frac{1.58 \times 10^{13}W}{1.74 \times 10^{17}W} \sim 10^{-4}. \qquad (46.3)$$

That is, mankind's energy consumption is comparable to fluctuations in the sun's energy output or, more accurately, the solar irradiance (i.e., including changes due to the earth's orbital parameters). So if we believe, as some do, that changes in insolation (solar radiation received by the earth) due to changes in the earth's orbital parameters are responsible for the ice ages that the earth has experienced over the last several million years, then it is just a short additional step to believe that twentieth-century industry's contribution to the earth's energy budget could also be responsible for dramatic earth climate changes.

Direct conversion of solar radiation to electrical energy has two benefits that are worth mentioning explicitly.

1. The fusion process ($4p^+ \rightarrow {}^4He^{++} + 2e^+ + 2\nu_e + 2\gamma$) that liberates energy also produces radioactivity in the surrounding environment. If we were able to control fusion on the earth's surface, we would still face the problem of caring for the nearby material made radioactive by this process. As it is, any radioactivity produced is produced in the sun. We do not have to worry about it.

2. Direct conversion of sunlight to electrical energy is a "neutral" process. It neither adds to nor subtracts energy from the earth's natural budget (in first order). At most, it redistributes energy from one geographical location to another. Because of this redistribution, there may be a second-order effect on the earth's energy budget. In short, this process is nonpolluting but could have local environmental impacts.

46.3 SIMPLE SOLAR CELLS

The spectrum of visible (to humans) solar radiation extends from the violet (400 nm or 4000 Å) to the red (700 nm or 7000 Å). The peak intensity is in the yellow at about 580 nm or 5800 Å. The sun behaves to a reasonable approximation as a blackbody with a temperature about 5778 K. (This blackbody looks yellow!)

Radiation in the violet at a wavelength of 400 nm consists of photons, each carrying an energy of about 3.1 eV. Similarly, "red photons" at 700 nm have an energy of 1.8 eV. Yellow photons carry 2.1 eV. We therefore expect solid-state devices that are designed to absorb solar radiation and convert it into electrical energy to have band gaps in the range of 1 to 3 eV.

A relatively simple solar conversion device is shown in Fig. 46.1. In this device two layers of impurity-doped silicon are placed over a glass substrate. One layer is doped with electron donors (*n* type), the other is doped with electron acceptors (*p* type). Both layers are bound to conducting metal contacts (the external leads, or wires). An antireflection coating is applied to the top of the cell to reduce losses of incident solar radiation by suppressing reflection. The glass substrate is usually designed to be reflective. This reflects unabsorbed solar radiation back through the cell, so the cell is effectively "twice as thick."

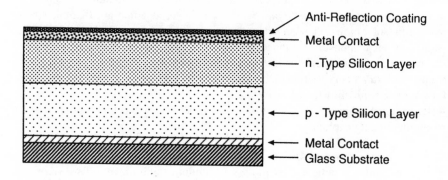

Fig. 46.1 Simple solar cell. n- and p-doped silicon layers are sandwiched between metallic contacts. The top of the cell is coated with an antireflection coating, and the glass substrate is usually reflective. Incident light creates an electron-hole pair. The electron is attracted to the n-type silicon layer, the hole is attracted to the p-type silicon layer.

The energy band structure for this device is shown schematically in Fig. 46.2. Two energy bands are shown. At zero temperature the lower (valence) band is filled and the upper (conduction) band is empty. At finite temperature some electrons are excited from the valence band to the conduction band. The probability that a state with energy E is filled is given by the Fermi-Dirac function

$$FD(E,T) = \frac{1}{e^{(E-\mu)/kT} + 1}. \tag{46.4}$$

Here T is the temperature, measured in kelvins, k is Boltzmann's constant, $k = 1.38 \times 10^{-16}$ erg/K $= 8.616 \times 10^{-5}$ eV/K, kT is an energy, and μ is a chemical potential. At room temperature ($T = 300$ K) $kT = 0.0258 \sim \frac{1}{40}$ eV.

Under equilibrium conditions, the probability that an electron is in any state in the conduction or valence band is determined by a single chemical potential. However, the absorption problem is not an equilibrium problem. Rather, it is a dynamic process for which we seek a steady state solution. Under this condition the probability distribution for electrons in the valence band is determined by one chemical potential, μ_V, while that for the conduction band is determined by another chemical potential, μ_C:

$$FD_{VB}(E,T) = \frac{1}{e^{(E-\mu_V)/kT} + 1}, \tag{46.5}$$

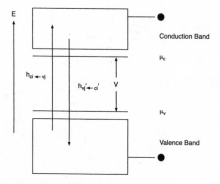

Fig. 46.2 Energy band diagram for a simple solar cell operating with two active bands, a conduction band and a valence band. The chemical potentials μ_C and μ_V, defining the nonequilibrium steady state populations in the two bands are shown.

$$FD_{CB}(E,T) = \frac{1}{e^{(E-\mu_C)/kT} + 1}. \qquad (46.6)$$

The potential difference, V, at which the cell operates, is determined by

$$qV = \mu_C - \mu_V. \qquad (46.7)$$

We now define $n(\nu, z)$ to be the number of photons of frequency ν that occur a distance z inside the cell ($z = 0$: front surface; $z = 1$: back surface). This number changes as we progress through the crystal for two reasons:

1. Photons of frequency ν are absorbed.

2. Photons of frequency ν are reemitted.

In the absorption process, an electron is removed from state j in the valence band and deposited in state i in the conduction band. The rate at which this happens is proportional to

1. the probability that state V_j is occupied: $FD_{V_j}(E,T)$

2. the probability that state C_i is not occupied: $1 - FD_{C_i}(E,T)$

3. a transition matrix element: $H_{C_i \leftarrow V_j}$

4. the number of photons: $n(\nu, z)$

5. the factor n/c, $n = $ index of refraction.

Absorption : $$\frac{dn(\nu, z)}{dz} = -\frac{n}{c} \sum_{i,j} H_{C_i \leftarrow V_j} \, n(\nu, z) FD_{V_j} (1 - FD_{C_i}) .$$

(46.8)

Photons are also reemitted into the field according to

Emission : $$\frac{dn(\nu, z)}{dz} = +\frac{n}{c} \sum_{i,j} H_{V_j \leftarrow C_i} \left[n(\nu, z) + 1 \right] FD_{C_i} (1 - FD_{V_j}) .$$

(46.9)

The emission contribution comes from two processes: stimulated emission (proportional to $n(\nu, z)$), the process responsible for laser action; and spontaneous emission (proportional to 1 within the square brackets). The matrix elements for the absorption and emission processes are complex conjugates of each other: $H_{C_i \leftarrow V_j} = H^*_{V_j \leftarrow C_i}$.

By integrating these equations through the crystal, from the front to the back, and then back to the front again, and imposing suitable boundary conditions, it is possible to compute qV (the energy delivered), the power output, and the operating efficiency of the solar cell, as a function of the band gap. This was done by Shockley and Queisser in 1961.[3] The result is that at room temperature, the theoretical maximum conversion efficiency is 40.7%. This occurs with a band gap of 1.1 eV.

Increasing the efficiency of a solar cell means that a smaller geographic area is required to produce an equivalent amount of electrical energy. To be explicit, the world's energy could be supplied by about 130,000 km^2 solar cells on the earth surface operating at 50% efficiency. By increasing the conversion efficiency by 1% we could produce the same energy with about 3000 km^2 smaller area.

Luque and Martí[4] have explored the possibility of increasing the conversion efficiency of a solar cell by inserting an impurity band between the valence and conduction bands (Fig. 46.3). They set up equations describing absorption and reemission of photons between all three pairs of bands and explored the energy band parameters that maximized the conversion efficiency of incident solar radiation to output electrical energy. In this configuration, the maximum theoretical conversion efficiency is 63.1%. This occurs when the band gap between the valence and conduction bands is 1.93 eV and the impurity band is 0.7 eV above the top of the valence band.

46.4 THE DESIGN PROBLEM

We now describe how to design a quantum mechanical device that approximately meets these specifications.

---

[3]W. Shockley and H. J. Queisser, "Detailed Balance Limit of Efficiency of p-n Junction Solar Cells," *Journal of Applied Physics* 32 (1961): 510-519.
[4]A. Luque and A. Martí, "Increasing the Efficiency of Ideal Solar Cells by Photon Induced Transitions at Intermediate Levels," *Physical Review Letters* 78 (1997): 5014-5017.

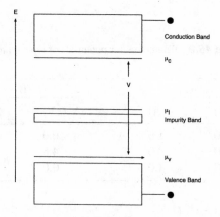

Fig. 46.3   Energy band diagram for a solar cell operating with three active bands: a conduction band, a valence band, and an impurity band between the two. The chemical potentials defining the nonequilibrium steady state populations in the three bands, μ_C, μ_I, and μ_V, are shown.

The steps involved in designing a device meeting the specifications just proposed— a gap of 1.93 eV between the valence band and the conduction band and an impurity band 0.7 eV above the valence band—are relatively simple. As a first step, we look for an "atom" whose lowest unoccupied state is more than 1.93 eV above the highest occupied state. When many atoms of this type are brought together into a regular lattice, the bound states are spread out into bands, and the energy between the top of the highest occupied (valence) band the the bottom of the lowest unoccupied or unfilled (conduction) band is reduced. For such substances, we compute the band structure. Those atoms ("A" atoms) for which the band gap is near 1.93 eV are of interest. We then search for atoms ("B" atoms) for which there is an unoccupied orbital 0.7 eV above the top of the valence band. Such atoms can serve as impurity atoms. When a lattice of A atoms is doped with B atoms, an impurity band will be produced between the A-type valence and conduction bands. If the doping percentage is small, the impurity band will be very narrow and the locations of the A-atom valence and conduction bands will be essentially unaffected.

To illustrate this process, we consider an A "atom" represented by a well of depth 7.0 eV and width 8.0 Å. Such atoms have four bound states at $E = -6.58$, -5.35, -3.37, and -0.87 eV, where we have chosen the potential at $\pm\infty$ to be 0.0 eV. If the lowest three levels are occupied, then the energy gap between the highest occupied level at -3.37 eV and the lowest unoccupied level at -0.87 eV is 2.5 eV. If the unit cell for the A atom in a lattice is described by potentials and widths as shown in Table 46.2, then each of the four levels is broadened into a band, as shown in Table 46.3. This table shows that the gap between the third (valence) and the fourth (conduction) band is 1.93 eV.

Table 46.2 Parameters for potentials of type A and B

A		B	
V (eV)	$\delta(\mathring{A})$	V (eV)	$\delta(\mathring{A})$
0.0	1.0	0.0	1.0
−7.0	8.0	−4.8	6.0
0.0	1.0	0.0	1.0

Table 46.3 Bound states and bands for atoms of type A

Level Number	Single Particle Energy	Top of Band Bottom of Band	Band Width	Band Gap
4	−0.87	−0.16 −1.19	1.03	
				1.93
3	−3.37	−3.12 −3.57	0.45	
				1.68
2	−5.35	−5.25 −5.42	0.17	
				1.14
1	−6.58	−6.56 −6.60	0.04	

With A as host atom, an atom of type B should have an unoccupied level at $-3.12 + 0.70 = -2.42$ eV. The potential B shown in Table 46.2 has bound states at $E = -4.19, -2.44$, and -0.14 eV. If the second level at -2.44 eV is unoccupied, B atoms will serve as impurity atoms in a lattice of A-type atoms to achieve the desired specifications. In Fig. 46.4 we show part of a one-dimensional lattice of A-type atoms with a single impurity atom of type B. Above this potential we show the four bands that arise from the four levels that the potential A possesses, as well as the three levels (dashed lines) provided by B atoms. As long as the doping level is small (few percent or less) the location of the bands and impurity levels is unaffected by the doping level—it is only the number of impurity states that changes with the doping density.

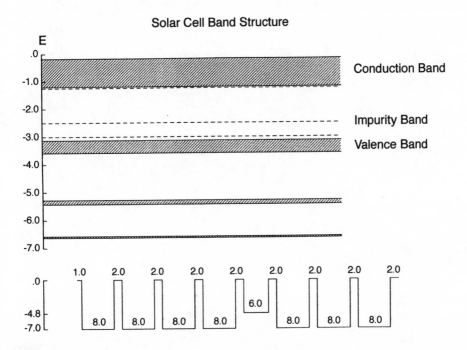

Fig. 46.4 A lattice of A atoms with a low density of B atoms produces a band structure as shown. Chemical potentials for the three bands are indicated by dashed lines. The potential parameters for the two "atoms" are presented in Table 46.2. This impurity-doped lattice meets the design criteria for a solar cell with maximum possible efficiency under the conditions stated.

In the real world we cannot design atomic potentials: we are stuck with what nature gives us. However, in the real world the basic ideas for designing devices are not much different from those described above.

Index